Applied Machine Learning Explainability Techniques

Make ML models explainable and trustworthy for practical applications using LIME, SHAP, and more

Aditya Bhattacharya

BIRMINGHAM—MUMBAI

Applied Machine Learning Explainability Techniques

Publishing Product Manager: Dinesh Chaudhary
Senior Editor: Tazeen Shaikh
Content Development Editor: Manikandan Kurup
Technical Editor: Devanshi Ayare
Copy Editor: Safis Editing
Project Coordinator: Farheen Fathima
Proofreader: Safis Editing
Indexer: Sejal Dsilva
Production Designer: Jyoti Chauhan
Marketing Coordinator: Shifa Ansari and Abeer Riyaz Dawe

First published: July 2022

Production reference: 2010722

Published by Packt Publishing Ltd.
Livery Place
35 Livery Street
Birmingham
B3 2PB, UK.

ISBN 978-1-80324-615-4

www.packt.com

To my lovely wife, Shreya. Thank you for being my eternal partner of our fairy tale called Life!

To my mother, Banani, and my father, Asit. Thank you for all your support, sacrifices and for teaching me the best lessons of life!

To my wonderful sister, Anuradha. Thank you for being my greatest advocate forever!

– Aditya Bhattacharya

Contributors

About the author

Aditya Bhattacharya is an explainable AI researcher at KU Leuven with 7 years of experience in data science, machine learning, IoT, and software engineering. Prior to his current role, Aditya worked in various roles in organizations such as West Pharma, Microsoft, and Intel to democratize AI adoption for industrial solutions. As the AI lead at West Pharma, he contributed to forming the AI Center of Excellence, managing and leading a global team of 10+ members focused on building AI products. He also holds a master's degree from Georgia Tech in computer science with machine learning and a bachelor's degree from VIT University in ECE. Aditya is passionate about bringing AI closer to end users through his various initiatives for the AI community.

I am immensely grateful to all who have been close to me and have supported me throughout the journey of writing this book, especially my wife Shreya, parents, sister, and all my aunts. A special thanks to my colleagues at the Augment research group of KU Leuven. I am grateful to Dr. Katrien Verbert for giving me the opportunity to pursue my journey as an XAI researcher. A big shoutout to all the reviewers for helping me throughout the writing process. Last, but not least, thanks to Abhijit Jana, who has always inspired me to step outside my comfort zone and pursue bigger challenges in life.

About the reviewers

Sumedh Vilas Datar is a machine learning engineer with 6 years of work experience in the field of deep learning, machine learning, and software engineering. He has a proven track record of single-handedly delivering end-to-end engineering solutions to real-world problems. He works at the intersection of engineering and products and has developed deep learning products from scratch that have been used by a lot of customers. Currently, Sumedh works in R&D on applied deep learning and has several granted patents and several more applied for. Sumedh studied biomedical engineering focused on computer vision and then went on to pursue a master's in computer science focused on AI.

Abhijit Jana is a trusted technology leader and advisor with 15 years of experience in the IT industry, with expertise in development, architecting, engineering, consulting, service delivery, and leadership. He is currently associated with West Pharmaceutical Services as a director of software engineering and is responsible for building and leading the software engineering team. Previously, he worked at Microsoft and is a former Microsoft MVP and Code Project MVP, and has been a speaker at various technology conferences. He is the author of the book *Kinect for Windows SDK Programming Guide* and coauthored the book *HoloLens Blueprint*. He is also the founder of Daily .NET Tips, a well-known website for developers, architects, and consultants.

Sk Nishan Ali works as a full stack data scientist at UnitedHealth Group. He has 6 years of experience across diverse areas of AI and machine learning, which include computer vision, classical machine learning, and natural language processing, and touches upon several business domains: healthcare, CRM, and sourcing. He has been instrumental in building high-performance end-to-end AI/ML products.

Table of Contents

3

Data-Centric Approaches

Section 2 – Practical Problem Solving

4

LIME for Model Interpretability

5
Practical Exposure to Using LIME in ML

6
Model Interpretability Using SHAP

7

Practical Exposure to Using SHAP in ML

8

Human-Friendly Explanations with TCAV

9

Other Popular XAI Frameworks

Section 3 – Taking XAI to the Next Level

10

XAI Industry Best Practices

Preface

Explainable AI (XAI) is an emerging field for bringing **artificial intelligence (AI)** closer to non-technical end-users. XAI promises to make **machine learning (ML)** models transparent, and trustworthy and promote AI adoption for industrial and research use-cases.

This book is designed with a unique blend of industrial and academic research perspectives for gaining practical skills in XAI. ML/AI experts working with data science, ML, deep learning, and AI will be able to put their knowledge to work with this practical guide to XAI for bridging the gap between AI and the end-user. The book provides a hands-on approach for implementation and associated methodologies of XAI that will have you up-and-running, and productive in no time.

Initially, you will get a conceptual understanding of XAI and why it's needed. Then, you will get the necessary practical experience of utilizing XAI in the AI/ML problem-solving process by making use of state-of-the-art methods and frameworks. Finally, you will get the necessary guidelines to take XAI to the next step and bridge the existing gaps between AI and end-users.

By the end of this book, you will be able to implement XAI methods and approaches using Python to solve industrial problems, address the key pain points encountered, and follow the best practices in the AI/ML life cycle.

Who this book is for

This book is designed for scientists, researchers, engineers, architects, and managers who are actively engaged in the field of ML and related areas. In general, anyone who is interested in problem-solving using AI would benefit from this book. You are recommended to have a foundational knowledge of Python, ML, deep learning, and data science. This book is ideal for readers who are working in the following roles:

- Data and AI scientists
- AI/ML engineers
- AI/ML product managers

- AI product owners

- AI/ML researchers

- User experience and HCI researchers

In general, any ML enthusiast with a foundational knowledge of Python will be able to read, understand and apply knowledge gained from this book.

What this book covers

Chapter 1, Foundational Concepts of Explainability Techniques, gives the necessary exposure to Explainable AI and help you understand it's importance. This chapter covers various terminology and concepts related to explainability techniques, which is frequently used throughout this book. This chapter also covers the key criteria of human-friendly explainable ML systems and different approaches to evaluating the quality of the explainability techniques.

Chapter 2, Model Explainability Methods, discusses the various model explainability methods used for explaining black-box models. Some of these are model agnostic, some are model specific. Some of these methods provide global interpretability while others provide local interpretability. This chapter will introduce you to a variety of techniques that can be used for explaining ML models and provides recommendation for the right choice of explainability method.

Chapter 3, Data-Centric Approaches, introduces the concept of data-centric XAI. This chapter covers various techniques to explain the working of ML systems in terms of the properties of the data, data volume, data consistency, data purity and actionable insights generated from the underlying training dataset.

Chapter 4, LIME for Model Interpretability, covers the application of one of the most popular XAI frameworks, called LIME. This chapter discusses about the intuition behind the working of the LIME algorithm and some important properties of the algorithm which makes the generated explanations human-friendly. Certain advantages and limitations of the LIME algorithm are also discussed in this chapter, along with a code tutorial for applying LIME for a classification problem.

Chapter 5, Practical Exposure to Using LIME in ML is an extension of the previous chapter, but more focused towards the practical applications of the LIME Python framework on different types of datasets like images, texts along with structured tabular data. Practical code examples are also covered in this chapter for providing exposure to on-hand knowledge using Python LIME framework. This chapter also covers if LIME is a good fit for production-level ML systems.

Chapter 6, Model Interpretability Using SHAP focuses on understanding the importance of the SHAP Python framework for model explainability. It covers the intuitive understanding of Shapley values and SHAP. This chapter also discusses how to use SHAP for model explainability through a variety of visualization and explainer methods. A code walkthrough for using SHAP to explain regression models is also covered in this chapter. Finally, we will discuss the key advantages and limitations of SHAP.

Chapter 7, Practical Exposure to Using SHAP in ML provides the necessary practical exposure of using SHAP with tabular structured data as well unstructured data like images and texts. We have discussed about the different explainers available in SHAP for both model-specific and model agnostic explainability. We have applied SHAP for explaining linear models, tree ensemble models, convolution neural network models and even transformer models in this chapter. Necessary code tutorials are also covered in this chapter for providing exposure to hands-on knowledge using Python SHAP framework.

Chapter 8, Human-Friendly Explanations with TCAV covers the concepts of TCAV, a framework developed by Google AI. This chapter provides both conceptual understanding of TCAV and practical exposure to applying the Python TCAV framework. The key advantages and limitations of TCAV are discussed along with interesting ideas about potential research problems that can be solved using concept-based explanations are discussed in the chapter.

Chapter 9, Other Popular XAI Frameworks covers about seven popular XAI frameworks available in Python – DALEX, Explainerdashboard, InterpretML, ALIBI, DiCE, ELI5, and H2O AutoML explainers. We have discussed about the supported explanation methods for each of the framework, practical application, and the various pros and cons of each framework. This chapter also provides a quick comparison guide for helping you decide which framework you should go for considering your own use-case.

Chapter 10, XAI Industry Best Practices focuses on the best practices for designing explainable AI systems for industrial problems. In this chapter, we have discussed about the open challenges of XAI and necessary design guidelines for explainable ML systems, considering the open challenges. We have also highlighted the importance of considering data-centric approaches of explainability, interactive machine learning and prescriptive insights for designing explainable AI/ML systems.

Chapter 11, End User-Centered Artificial Intelligence introduces the ideology of end user centered artificial intelligence (ENDURANCE) for the design and development of explainable AI/ML Systems. We have discussed about the importance of using XAI to steer towards the main goals of the end user for building explainable AI/ML systems. Using some of principles and recommended best practices presented in the chapter, we can bridge the gap between AI and the end user to a great extent!

To get the most out of this book

To run the code tutorials provided in this book, you will need a Jupyter environment with Python 3.6+. This can be achieved in either of the following ways:

- Install one on your machine locally via **Anaconda Navigator** or from scratch with **pip**.

- Use a cloud-based environment such as **Google Colaboratory**, **Kaggle notebooks**, **Azure notebooks**, or **Amazon SageMaker**.

You can take a look at the supplementary information provided at the code repository if you are new to Jupyter notebooks: `https://github.com/PacktPublishing/Applied-Machine-Learning-Explainability-Techniques/blob/main/SupplementaryInfo/CodeSetup.md`.

You can also take a look at `https://github.com/PacktPublishing/Applied-Machine-Learning-Explainability-Techniques/blob/main/SupplementaryInfo/PythonPackageInfo.md` and `https://github.com/PacktPublishing/Applied-Machine-Learning-Explainability-Techniques/blob/main/SupplementaryInfo/DatasetInfo.md` for getting the supplementary information about the Python packages and datasets used in the tutorial notebooks.

For instructions on installing the Python packages used throughout the book, please refer the specific notebook provided in the code repository. For any additional help needed, please refer the original project repository of the specific package. You can use **PyPi** (`https://pypi.org/`) and search for the specific package and navigate to the code repository of the project. It is expected that installation or execution instructions of these packages can change from time to time, given how often packages change. We also tested the code with specific versions detailed in the *Python package information README file* under the supplementary information provided at the code repository. So, if anything doesn't work as expected with the later versions, please install the specific version mentioned in the README instead.

If you are using the digital version of this book, we advise you to type the code yourself or access the code from the book's GitHub repository (a link is available in the next section). Doing so will help you avoid any potential errors related to the copying and pasting of code.

For beginners without any exposure to ML or data science, it is recommended to read the book sequentially as many important concepts are explained in sufficient detail in the earlier chapters. Seasoned ML or data science experts who are relatively new to the field of XAI can skim through the first three chapters to get clear conceptual understanding of various terminology used. For chapters four to nine, any order should be fine for seasoned experts. For all level of practitioners, it is recommended that you read chapter 10 and 11 only after covering all the nine chapters.

Regarding the code provided, it is recommended that you either read each chapter and then run the corresponding code, or you can run the code simultaneously while reading the specific chapters. Sufficient theory is also added in the Jupyter notebooks to help you understand the overall flow of the notebook.

When you are reading the book, it is recommended that you take notes of the important terminologies covered and try to think of ways in which you could apply the concept or the framework learned. After reading the book and going through all the Jupyter notebooks, hopefully, you will be inspired to apply the newly gained knowledge into action!

Download the example code files

You can download the example code files for this book from GitHub at https://github.com/PacktPublishing/Applied-Machine-Learning-Explainability-Techniques. If there's an update to the code, it will be updated in the GitHub repository.

We also have other code bundles from our rich catalog of books and videos available at https://github.com/PacktPublishing/. Check them out!

Download the color images

We also provide a PDF file that has color images of the screenshots and diagrams used in this book. You can download it here: https://packt.link/DF71G.

Conventions used

There are a number of text conventions used throughout this book.

`Code in text`: Indicates code words in text, database table names, folder names, filenames, file extensions, pathnames, dummy URLs, user input, and Twitter handles. Here is an example: "For this example, we will use the `RegressionExplainer` and `ExplainerDashboard` submodules."

A block of code is set as follows:

```
pdp = PartialDependence(
    predict_fn=model.predict_proba,
    data=x_train.astype('float').values,
    feature_names=list(x_train.columns),
    feature_types=feature_types)
pdp_global=pdp.explain_global(name='Partial Dependence')
```

When we wish to draw your attention to a particular part of a code block, the relevant lines or items are set in bold:

```
explainer = shap.Explainer(model, x_test)
shap_values = explainer(x_test)
shap.plots.waterfall(shap_values[0], max_display = 12,
                     show=False)
```

Bold: Indicates a new term, an important word, or words that you see onscreen. For instance, words in menus or dialog boxes appear in **bold**. Here is an example: "Due to these known drawbacks, the search for a robust **Explainable AI (XAI)** framework is still on."

> Tips or important notes
> Appear like this.

Get in touch

Feedback from our readers is always welcome.

General feedback: If you have questions about any aspect of this book, email us at `customercare@packtpub.com` and mention the book title in the subject of your message.

Errata: Although we have taken every care to ensure the accuracy of our content, mistakes do happen. If you have found a mistake in this book, we would be grateful if you would report this to us. Please visit www.packtpub.com/support/errata and fill in the form.

Piracy: If you come across any illegal copies of our works in any form on the internet, we would be grateful if you would provide us with the location address or website name. Please contact us at copyright@packt.com with a link to the material.

If you are interested in becoming an author: If there is a topic that you have expertise in and you are interested in either writing or contributing to a book, please visit authors.packtpub.com.

Share Your Thoughts

Once you've read *Applied Machine Learning Explainability Techniques*, we'd love to hear your thoughts! Scan the QR code below to go straight to the Amazon review page for this book and share your feedback.

https://packt.link/r/1803246154

Your review is important to us and the tech community and will help us make sure we're delivering excellent quality content.

Section 1 – Conceptual Exposure

This section will give you the necessary conceptual exposure to explainability techniques for **machine learning** (**ML**) models with practical examples. You will learn about the foundational concepts, different dimensions of explainability, various model explainability methods, and even data-centric approaches to explainability. Knowledge of the foundational concepts will help you understand the guidelines for designing robust explainable ML systems like those covered in this book.

This section comprises the following chapters:

- *Chapter 1, Foundational Concepts of Explainability Techniques*
- *Chapter 2, Model Explainability Methods*
- *Chapter 3, Data-Centric Approaches*

1
Foundational Concepts of Explainability Techniques

As more and more organizations have started to adopt **Artificial Intelligence** (**AI**) and **Machine Learning** (**ML**) for their critical business decision-making process, it becomes an immediate expectation to interpret and demystify **black-box algorithms** to increase their adoption. AI and ML are being increasingly utilized for determining our day-to-day experiences across multiple areas, such as banking, healthcare, education, recruitment, transport, and supply chain. But the integral role played by AI and ML models has led to the growing concern of business stakeholders and consumers about the lack of transparency and interpretability as these black-box algorithms are highly subjected to human bias; particularly for high-stake domains, such as healthcare, finance, legal, and other critical industrial operations, model explainability is a prerequisite.

As the benefits of AI and ML can be significant, the question is, how can we increase its adoption despite the growing concerns? Can we even address these concerns and democratize the use of AI and ML? And how can we make AI more explainable for critical industrial applications in which black-box models are not trusted? Throughout this book, we will try to learn the answers to these questions and apply these concepts and ideas to solve practical problems!

In this chapter, you will learn about the foundational concepts of **Explainable AI (XAI)** so that the terms and concepts used in future chapters are clear, and it will be easier to follow and implement some of the advanced explainability techniques discussed later in this book. This will give you the required theoretical knowledge needed to understand and implement the practical techniques discussed in later chapters. The chapter focuses on the following main topics:

- Introduction to XAI
- Defining explanation methods and approaches
- Evaluating the quality of explainability methods

Now, let's get started!

Introduction to XAI

XAI is the most effective practice to ensure that AI and ML solutions are transparent, trustworthy, responsible, and ethical so that all regulatory requirements on algorithmic transparency, risk mitigation, and a fallback plan are addressed efficiently. AI and ML explainability techniques provide the necessary visibility into how these algorithms operate at every stage of their solution life cycle, allowing end users to understand *why* and *how* queries are related to the outcome of AI and ML models.

Understanding the key terms

Usually, for ML models, for addressing the *how* questions, we use the term *interpretability*, and for addressing the *why* questions, we use the term *explainability*. In this book, the terms **model interpretability** and **model explainability** are interchangeably used. However, for providing *human-friendly* holistic explanations of the outcome of ML models, we will need to make ML algorithms both interpretable and explainable, thus allowing the end users to easily comprehend the decision-making process of these models.

In most scenarios, ML models are considered as black-boxes, where we feed in any training data and it is expected to predict on new, unseen data. Unlike conventional programming, where we program specific instructions, an ML model automatically tries to learn these instructions from the data. As illustrated in *Figure 1.1*, when we try to find out the rationale for the model prediction, we do not get enough information!

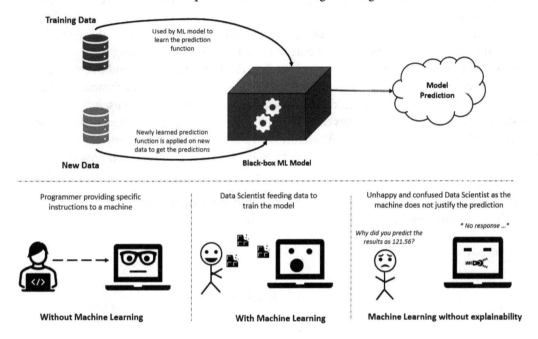

Figure 1.1 – Conventionally, black-box models do not provide any rationale behind predictions

Now, let's understand the impact of incorrect predictions and inaccurate ML models.

Consequences of poor predictions

Traditionally, all ML models were believed to be magical black-boxes that can automatically decipher interesting patterns and insights from the data and provide *silver bullet* outcomes! As compared to conventional rule-based computer programs, which are limited by the intelligence of the programmer, well-trained ML algorithms are considered to provide rich insights and accurate predictions even in complex situations. But the fact is, all ML models suffer from *bias*, which can be due to the **inductive bias** of the algorithm itself, or it can be due to the presence of bias in the dataset used for training the model. In practice, there can be other reasons, such as **data drift**, **concept drift**, and **overfitted** or **underfitted** models, for which model predictions can go wrong. As the famous British statistician George E.P. Box once said, *"All models are wrong, but some are useful"*; all statistical, scientific, and ML models can give incorrect outcomes if the initial assumptions of these methods are not consistent. Therefore, it is important for us to know why an ML model predicted a specific outcome, what can be done if it is wrong, and how the predictions can be improved.

Figure 1.2 illustrates a collection of news headlines highlighting the failure of AI algorithms towards producing fair and unbiased outcomes.

Figure 1.2 – Growing concern of bias and lack of fairness of ML models being reported frequently

Before completely agreeing with me on the necessity of model explainability, let me try to give some practical examples of low-stake and high-stake domains to understand the consequences of poor predictions. Weather forecasting is one of the classical forecasting problems that is extremely challenging (as it depends on multiple dynamic factors) where ML is extensively used, and the ability of ML algorithms to consider multiple parameters of different types makes it more efficient than standard statistical models to predict the weather. Despite having highly accurate forecast models, there are times when weather forecasting algorithms might miss the prediction of rainfall, even though it starts raining after a few minutes! But the consequences of such a poor prediction might not be so severe, and moreover, most people do not blindly rely on automated weather predictions, thus making weather forecasting a low-stake domain problem.

Similarly, for another low-stake domain, such as a content recommendation system, even if an ML algorithm provides an irrelevant recommendation, at the most, the end users might spend more time explicitly searching for relevant content. While the overall experience of the end user might be impacted, still, there is no severe loss of life or livelihood. Hence, the need for model explainability is not critical for low-stake domains, but providing explainability to model predictions does make the automated intelligent systems more trustworthy and reliable for end users, thus increasing AI adoption by enhancing the end user experience.

Now, let me give an example where the consequences of poor predictions led to a severe loss of reputation and valuation of a company, impacting many lives! In November 2021, an American online real estate marketplace company called *Zillow* (https://www.zillow.com/) reported having lost over 40% of its stock value, and the home-buying division *Offers* lost over $300 million because of its failure to detect the unpredictability of their home price forecasting algorithms (for more information, please refer to the sources mentioned in the *References* section). In order to compensate for the loss, Zillow had to take drastic measures of cutting down its workforce and several thousands of families were impacted.

Similarly, multiple technology companies have been accused of using highly biased AI algorithms that could result in social unrest due to racial or gender discrimination. One such incident happened in 2015 when Google Photos made a massive racist blunder by automatically tagging an African-American couple as *Gorilla* (please look into the sources mentioned in the *References* section for more information). Although these blunders were unintentional and mostly due to biased datasets or non-generalized ML models, the consequences of these incidents can create massive social, economic, and political havoc. Bias in ML models in other high-stake domains, such as healthcare, credit lending, and recruitment, continuously reminds us of the need for more transparent solutions and XAI solutions on which end users can rely.

As illustrated in *Figure 1.3*, the consequences of poor predictions highlight the importance of XAI, which can provide early indicators to prevent loss of reputation, money, life, or livelihood due to the failure of AI algorithms:

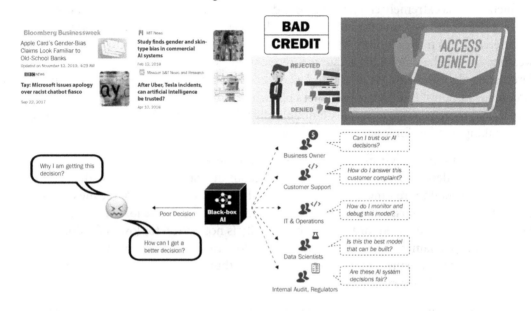

Figure 1.3 – Common consequences of poor prediction of ML models

Now, let's try to summarize the need for model explainability in the next section.

Summarizing the need for model explainability

In the previous section, we learned that the consequences of poor predictions can impact many lives in high-risk domains, and even in low-risk domains the end user's experience can be affected. *Samek and Binder's* work in *Tutorial on Interpretable Machine Learning, MICCAI'18*, highlights the main necessity of model explainability. Let me try to summarize the key reasons why model explainability is essential:

- **Verifying and debugging ML systems**: As we have seen some examples where wrong model decisions can be costly and dangerous, model explainability techniques help us to verify and validate ML systems. Having an interpretation for incorrect predictions helps us to debug the root cause and provides a direction to fix the problem. We will discuss the different stages of an explainable ML system in more detail in *Chapter 10, XAI Industry Best Practices*.

- **Using user-centric approaches to improve ML models**: XAI provides a mechanism to include human experience and intuition to improve ML systems. Traditionally, ML models are evaluated based on prediction error. Using such evaluation approaches to improve ML models doesn't add any transparency and may not be robust and efficient. However, using explainability approaches, we can use human experience to verify predictions and understand whether model-centric or data-centric approaches are further needed to improve the ML model. *Figure 1.4* compares a classical ML system with an explainable ML system:

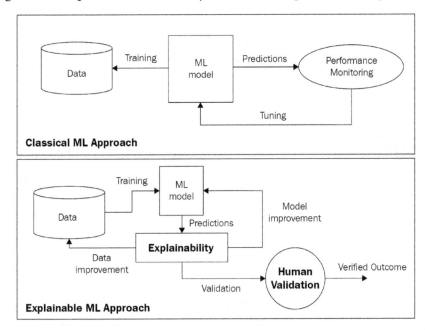

Figure 1.4 – Comparison between classical ML and explainable ML approach

- **Learning new insights**: ML is considered to automatically unravel interesting insights and patterns from data that are not obvious to human beings. Explainable ML provides us with a mechanism to understand the rationale behind the insights and patterns automatically picked up by the model and allows us to study these patterns in detail to make new discoveries.

- **Compliance with legislation**: Many regulatory bodies, such as the **General Data Protection Regulation (GDPR)** and the **California Consumer Privacy Act (CCPA)**, have expressed severe concerns about the lack of explainability in AI. So, growing global AI regulations have empowered individuals with the right to demand an explanation of automated decision-making systems that can affect them. Model explainability techniques try to ensure ML models are compliant with proposed regulatory laws, thereby promoting fairness, accountability, and transparency.

"

Companies should commit to ensuring systems that could fall under GDPR, including AI, will be complaint. The threat of **sizeable fines of €20 million or 4% of global turnover** provides a sharp incentive.

Article 22 of GDPR empowers individuals with the **right to demand an explanation of how an AI system made a decision that affects them.**

"

- European Commision

Andrus Ansip
@Ansip_EU

You have the right to be informed about an automated decision and ask for a human being to review it, for example if your online credit application is refused. #EUdataP #GDPR #AI #digitalrights #EUandMe europa.eu/!nN77Dd

#DIGITALRIGHTS
In the Digital Single Market
Stronger data protection including **rights** to
· be **forgotten**
· **move** your data
· **know** which data is collected about you, if your data has been leaked or hacked
· be informed about **automated decisions**

VP, European Commision

Regulatory Acts	Descriptions
GDPR	Article 22 empowers individuals with the right to demand an explaination of the decision-making process of automated systems
California Consumer Privacy Act 2019	Requires companies to align the process of collection, storage, and sharing of personal data with the new requirements of January 1, 2020
Algorithmic Accountability Act 2019	Mandates organizations to provide assessments of the risks of having automated decision systems to privacy, security, inaccurate, unfair, biased, or discriminatory outcomes impacting consumers.
Washington Bill 1655	Introduces measures for the use of automated decision systems to protect consumers, improve transparency and create more market predictability.
Illinois House Bill 3415	Establishes guidelines for not including information related to applicant race or zip code for predictive data analytics for the purpose of hiring or financial services
Massachusetts Bill H.2701	Establishes guidelines on automated decision-making, transparency, fairness, and individual rights.

Figure 1.5 – (Top) Screenshots of tweets from the European Commission, highlighting the right to demand explanations. (Bottom) Table showing some important regulatory laws established for making automated decision systems explainable, transparent, accountable, and fair

The need for model explainability can be visualized in the following diagram of the **FAT model of explainable ML** as provided in the book *Interpretable Machine Learning with Python* by *Serg Masís*.

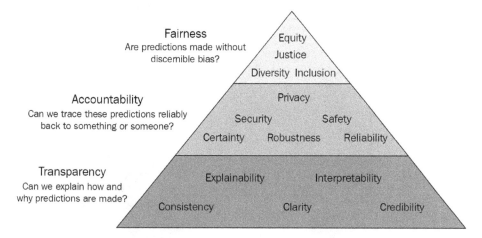

Figure 1.6 – FAT model of explainable ML (from Interpretable Machine Learning with Python by Serg Masís)

Figure 1.6 shows the pyramid that forms the FAT model of explainable ML system for increasing AI adoption. Let us discuss about defining explanation methods and approaches in the next section.

Defining explanation methods and approaches

In this section, let's try to understand some key concepts required for understanding and applying various explainability techniques and approaches.

Dimensions of explainability

Adding to the concepts presented at *MICCAI'18 from Tutorial on Interpretable Machine Learning* by *Samek and Binder*, when we talk about the problem of demystifying black-box algorithms, there are four different dimensions through which we can address this problem, as can be seen in the following diagram:

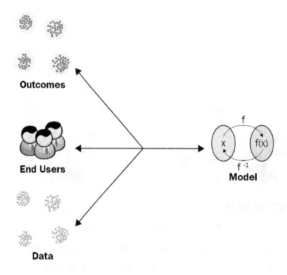

Figure 1.7 – Four dimensions of explainability

Now, let's learn about these dimensions in detail:

- **Data**: The dimension of explainability revolves around the underlying data that is being modeled. Understanding the data, identifying its limitations and relevant components, and forming certain hypotheses are crucial to setting up the correct expectations. A robust data curation process, analyzing data purity, and the impact of adversarial effects on the data are other key exercises done for obtaining explainable outcomes.

- **Model**: Model-based interpretability techniques often help us to understand how the input data is mapped to the output predictions and be aware of some limitations and assumptions of the ML algorithms used. For example, the Naïve Bayes algorithm used for ML classification assumes that the presence of a certain feature is completely independent and unrelated to the presence of any other features. So, knowing about these *inductive biases* of ML algorithms helps us to understand and anticipate any prediction error or limitations of the ML models.

- **Outcomes**: The outcome of explainability is about understanding *why* a certain prediction or decision is made by an ML model. Although data and model interpretability is quite crucial, most ML experts and end-users focus on making the final model predictions interpretable.

- **End users**: The final dimension of explainability is all about creating the right level of abstraction and including the right amount of details for the final consumers of the ML models so that the outcomes are reliable and trustworthy for any non-technical end-user and empower them to understand the decision-making process of black-box algorithms.

Explainability to AI/ML algorithms is provided with respect to one or more dimensions of explainability. Next, let's discuss about addressing the key questions of explainability.

Addressing key questions of explainability

Now that we understand the different dimensions of explainability, let's discuss what is needed to make ML models explainable. In order to make ML algorithms explainable, the following are the key questions that we should try to address:

- *What do we understand from the data?*

  The very first step is all about the data. Before even proceeding with the AI and ML modeling, we should spend enough time analyzing and exploring the data. The goal is always to look for gaps, inconsistencies, potential biases, or hypotheses that might impact or create challenges while modeling the data and generating the predictions. This helps us to know what is expected and how certain aspects of the data can contribute toward solving the business problem.

- *How is the model created?*

  We need to understand how transparent the algorithm is and what kind of relationship the algorithm can capture when modeling the data during the modeling process. This is the step where we try to understand the inductive bias of the algorithms and then try to relate this to the initial hypothesis or observations obtained while exploring the data. For example, linear models will not model the data efficiently if the data has some quadratic or cycle patterns observed using visualization-based data exploration methods. The prediction error is expected to be higher. So, if it is unclear how the algorithm builds a model of the training data, these algorithms are less transparent and, hence, less interpretable.

- *What do we know about the global interpretability of the trained model?*

   Understanding the global model interpretability is always challenging. It is about getting a holistic view of the underlying features used, knowing the important features, how sensitive the model is toward changes in the key feature values, and what kind of complex interactions are happening inside the model. This is especially hard to achieve in practice for complex deep learning models that have millions of parameters to learn and several hundreds of layers.

- *What is the influence of different parts of the model on the final prediction?*

   Different parts of an ML model might impact the final prediction in a different way. Especially for deep neural network models, each layer tries to learn different types of features. When model predictions are incorrect, understanding how different parts of a model can affect or control the final outcome is very important. So, explainability techniques can unravel insights from different parts of a model and help debug and observe the algorithm's robustness for different data points.

- *Why did the model make a specific prediction for a single record and a batch of records?*

   The most important aspect of explainability is understanding why the model is making a specific prediction and not something else. So, certain local and global explanation techniques are applied, which either consider the impact of individual features or even the collective impact of multiple features on the outcome. Usually, these explainability techniques are applied for single instances of the data and a batch of data instances to understand whether the observations are consistent.

- *Does the outcome match the expectation of the end user?*

   The final step is always providing user-centric explanations. This means explainability is all about comparing the outcome with end users' predictions based on common sense and human intuition. If the model forecast matches the user's prediction, providing a reasonable explanation includes justifying the dominant factors for the specific outcome. But suppose the model forecasting is not matching the user's prediction. In that case, a good explanation tries to justify what changes could have happened in the input observations to get a different outcome.

For example, let's say, considering usual weekday traffic congestion, the time taken to reach from office to home for me is 30 minutes. But if it is raining, I would expect the vehicles on the road to move slowly and traffic congestion to be higher, and hence might expect it to take longer to reach home. Now, if an AI application predicts the time to get home as still 30 minutes, I might not trust this prediction as this is counter-intuitive.

Now, let's say that the algorithm was accurate in its forecast. However, the justification provided to me was about the movement of the vehicles on my route, and the AI app just mentioned that the vehicles on my route are moving at the same speed as on other days. Does this explanation really help me to understand the model predictions? No, it doesn't. But suppose the application mentions that there are fewer vehicles on the route than found on typical days. In that case, I would easily understand that the number of vehicles is fewer due to the rain and hence the time to destination is still the same as usual on weekdays.

My own recommendation is that, after training and validating an ML model, always try to seek answers to these questions as an initial step in interpreting the working of black-box models.

Understanding different types of explanation methods

In the previous section, we discussed some key questions to address when designing and using robust explainability methods. In this section, we will discuss various types of explanation methods, considering the four dimensions of explainability used in ML:

- **Local explainability** and **global explainability**: ML model explainability can be done for single local instances of the data to understand how a certain range of values or specific categorical value can be related to the final prediction. This is called local explainability. Global model explainability is used to explain the behavior of the entire model or certain important features as a whole that contribute toward a specific set of model outcomes.

- **Intrinsic explainability** and **extrinsic explainability**: Some ML models, such as linear models, simple decision trees, and heuristic algorithms, are intrinsically explainable as we clearly know the logic or the mathematical mapping of the input and output that the algorithm applies, whereas extrinsic or post hoc explainability is about first training an ML model on given data and then using certain model explainability techniques separately to understand and interpret the model's outcome.

- **Model-specific explainability** and **model-agnostic explainability**: When we use certain explainability methods that are applicable for any specific algorithm, then these are model-specific approaches. For example, visualization of the tree structure in decision tree models is only specific to the decision tree algorithm and hence comes under the model-specific explainability method. Model-agnostic methods are used to provide explanations to any ML model irrespective of the algorithm being used. Mostly, these are post hoc analysis methods, used after the trained ML model is obtained, and usually, these methods are not aware of the internal model structure and weights. In this book, we will mostly focus on model-agnostic explainability methods, which are not dependent on any particular algorithm.

- **Model-centric explainability** and **data-centric explainability**: Conventionally, the majority of explanation methods are model-centric, as these methods try to interpret how the input features and target values are being modeled by the algorithm and how the specific outcomes are obtained. But with the latest advancement in the space of data-centric AI, ML experts and researchers are also investigating explanation methods around the data used for training the models, which are known as data-centric explainability. Data-centric methods are used to understand whether the data is consistent, well curated, and well suited for solving the underlying problem. Data profiling, detection of data and concept drifts, and adversarial robustness are certain specific data-centric explainability approaches that we will be discussing in more detail in *Chapter 3, Data-Centric Approaches*.

We will discuss all these types of explainability methods in later chapters of the book.

Understanding the accuracy interpretability trade-off

For an ideal scenario, we would want our ML models to be highly accurate and highly interpretable so that any non-technical business stakeholder or end user can understand the rationale behind the model predictions. But in practice, achieving highly accurate and interpretable models is extremely difficult, and there is always a trade-off between accuracy and interpretability.

For example, to perform radiographic image classification, intrinsically interpretable ML algorithms, such as decision trees, might not be able to give efficient and generalized results, whereas more complex deep convolutional neural networks, such as DenseNet, might be more efficient and robust for modeling radiographic image data. But DenseNet is not intrinsically interpretable, and explaining the algorithm's working to any non-technical end user can be pretty complicated and challenging. So, highly accurate models, such as deep neural networks, are non-linear and more complex and can capture complex relationships and patterns from the data, but achieving interpretability is difficult for these models. Highly interpretable models, such as linear regression and decision trees, are primarily linear and less complex, but these are limited to learning only linear or less-complex patterns from the data.

Now, the question is, is it better to go with highly accurate models or highly interpretable models? I would say that the correct answer is, *it depends!* It depends on the problem being solved and on the consumers of the model. For high-stake domains, where the consequences of poor predictions are severe, I would recommend going for more interpretable models even if accuracy is being sacrificed. Any rule-based heuristic model that is highly interpretable can be very effective in such situations. But if the problem is well studied, and getting the least prediction error is the main goal (such as in any academic use case or any ML competitions) such that the consequences of poor prediction will not create any significant damage, then going for highly accurate models can be preferred. In most industrial problems, it is essential to keep the right balance of model accuracy and interpretability to promote AI adoption.

Figure 1.10 illustrates the accuracy-interpretability trade-off of popular ML algorithms:

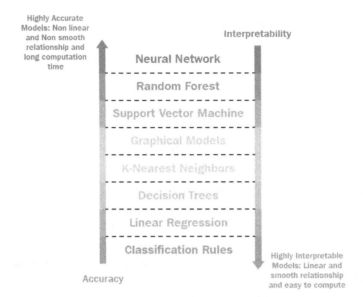

Figure 1.8 – Accuracy-interpretability trade-off diagram

Now that we have a fair idea of the accuracy-interpretability trade-off, let's try to understand how to evaluate the quality of explainability methods.

Evaluating the quality of explainability methods

Explainability is subjective and may vary from person to person. The key question is *How do we determine whether one approach is better than the other?* So, in this section, let's discuss certain criteria to consider for evaluating explainability techniques for ML systems.

Criteria for good explainable ML systems

As explainability for ML systems is a very subjective topic, first let's try to understand some key criteria for good human-friendly explanations. In his book *Interpretable Machine Learning*, *Christoph Molnar*, the author, has also tried to emphasize the importance of good human-friendly explanations after thorough research, which I will try to mention in a condensed form considering modern, industrial, explainable ML systems:

- **Coherence with a priori knowledge**: Consistency with prior beliefs of end users is an important criterion of explainable ML systems. If any explanation contradicts a priori knowledge of human beings, then humans tend to have less trust in such explanations. However, it is challenging to introduce prior knowledge of humans into ML models. But human-friendly explainable ML systems should try to provide explanations surrounding certain features that have direct and less complex relationships with the outcome, such that the relationship is coherent with the prior beliefs of the end users.

  For example, for predicting the presence of diabetes, the measure of blood glucose level has a direct relationship, which is consistent with prior human beliefs. If the blood glucose level is higher than usual, this might indicate diabetes for the patient, although diabetes can also be due to certain genetic factors or other reasons. Similarly, high blood glucose levels can also be momentary and a high blood glucose level doesn't always mean that the patient has diabetes. But as the explanation is consistent with a priori knowledge, end users will have more trust in such explanations.

- **Fidelity**: Another key factor for providing a holistic explanation is the truthfulness of the explanation, which is also termed the fidelity of ML models. Explanations with high fidelity can approximate the holistic prediction of the black-box models, whereas low-fidelity explanations can interpret a local data instance or a specific subset of the data. For example, for doing sales forecasting, providing explanations based on just the trend of historical data doesn't give a complete picture, as other factors, such as production capacity, market competition, and customer demand, might influence the outcome of the model. Fidelity plays a key role, especially for doing a detailed root cause analysis, but too many details may not be useful for common users, unless requested.

- **Abstraction**: Good human-friendly explanations are always expected in a concise and abstract format. Too many complicated details can also impact the experience of end users. For example, for weather forecasting, if the model predicts a high probability of rain, the concise and abstract explanation can be that it is cloudy now and raining within 5 kilometers of the current location, so there is a high probability of rain.

  But if the model includes details related to precipitation level, humidity, and wind speed, which might also be important for the prediction of rainfall, these additional details are complex and difficult to comprehend, and hence not human-friendly. So, good, human-friendly explanations include the appropriate amount of abstraction to simplify the understanding for the end users. End users mostly prefer concise explanations, but detailed explanations might be needed when doing root cause analysis for model predictions.

- **Contrastive explanations**: Good, human-friendly explanations are not about understanding the inner workings of the models but mostly about comparing the *what-if* scenarios. Suppose the outcome is continuous numerical data as in the case of regression problems. In that case, a good explanation for predictions includes comparing with another instance's prediction that is significantly higher or lower. Similarly, a good explanation for classification problems is about comparing the current prediction with other possible outcomes. But contrastive explanations are application-dependent as it requires a point of comparison, although understanding the *what-if* scenarios helps us to understand the importance of certain key features and how these features are related to the target variable.

  For example, for a use case of employee churn prediction, if the model predicts that the employee is likely to leave the organization, then contrastive explanations try to justify the model's decision by comparing it with an instance's prediction where the employee is expected to stay in the organization and compare the values of the key features used to model the data. So, the explanation method might convey that since the salary of the employee who is likely to leave the organization is much lower than that of employees who are likely to stay within the organization, the model predicted that the employee is expected to leave the organization.

- **Focusing on the abnormality**: This may sound counter-intuitive, but human beings try to seek explanations for events that are not expected and not obvious. Suppose there is an abnormal observation in the data, such as a rare categorical value or an anomalous continuous value that can influence the outcome. In that case, it should be included in the explanation. Even if other normal and consistent features have the same influence on the model outcome, still including the abnormality holds higher importance in terms of human-friendly explanation.

  For example, say we are predicting the price of cars based on their configuration, let's say the mode of operation is *electric*, which is a rare observation compared to *gasoline*. Both of these categories might have the same influence on the final model prediction. Still, the model explanation should include the rare observation, as end users are more interested in abnormal observations.

- **Social aspect**: The social aspect of model explainability determines the abstraction level and the content of explanations. The social aspect depends on the level of understanding of the specific target audience and might be difficult to generalize and introduce during the model explainability method. For example, suppose a stock-predicting ML model designed to prescribe actions to users suggests *shorting* a particular stock. In that case, end users outside the finance domain may find it difficult to understand. But instead, if the model suggests selling *a stock at the current price without possessing it and buying it back after 1 month when the price is expected to fall*, any non-technical user might comprehend the model suggestions easily. So, good explanations consider the social aspect, and often, user-centric design principles of **Human-Computer Interaction** (**HCI**) are utilized to design good, explainable ML systems that consider the social aspect.

Now that we have a fair idea of the key criteria for good explanations, in the next section, we will discuss some auxiliary criteria that are equally important while building explainable ML systems.

Auxiliary criteria of XAI for ML systems

Good explanations are not limited to the key criteria discussed previously, but there are a few auxiliary criteria of XAI as discussed by *Doshi-Velez* and *Kim* in their work *Towards A Rigorous Science of Interpretable Machine Learning*:

- **Unbiasedness**: Model explainability techniques should also look for the presence of any form of bias in the data or the model. So, one of the key goals of XAI is to make ML models unbiased and fair. For example, for predicting credit card fraud, the explainability approach should investigate the importance of demographic information related to the gender of the customer for the model's decision-making process. If the importance of gender information is high, that means that the model is biased toward a particular gender.

- **Privacy**: Explainability methods should comply with data privacy measures, and hence any sensitive information should not be used for the model explanations. Mainly for providing personalized explanations, ensuring compliance with data privacy can be very important.

- **Causality**: Model explainability approaches should try to look for any causal relationships so that the end users are aware that due to any perturbation, there can be changes in model predictions for production systems.

- **Robustness**: Methods such as sensitivity analysis help to understand how robust and consistent a model prediction is with respect to its feature values. If small changes in input features lead to a significant shift in model predictions, it shows that the model is not robust or stable.

- **Trust**: One of the key goals of XAI is to increase AI adoption by increasing the trust of the end users. So, all explainability methods should make black-box ML models more transparent and interpretable so that the end users can trust and rely on them. If explanation methods don't meet the criteria of good explanations, as discussed in the *Criteria for good explainable ML systems* section, it might not help to increase the trust of its consumers.

- **Usable**: XAI methods should try to make AI models more usable. Hence, it should provide information to the users to accomplish the task. For example, counterfactual explanations might suggest a loan applicant pays their credit card bill on time for the next 2 months and clear off their previous debts before applying for a new loan so that their loan application is not rejected.

Next, we will need to understand various levels of evaluating explainable ML systems.

Taxonomy of evaluation levels for explainable ML systems

Now that we have discussed the key criteria for designing and evaluating good explainable ML systems, let's discuss the taxonomy of evaluation methods for judging the quality of explanations. In their work *Towards A Rigorous Science of Interpretable Machine Learning*, *Doshi-Velez and Kim* mentioned three major types of evaluation approaches that we will try to understand in this section. Since explainable ML systems are to be designed with user-centric design principles of HCI, human beings evaluating real tasks play a central role in assessing the quality of explanation.

But human evaluation mechanisms can have their own challenges, such as different types of human bias and being more time- and other resource-consuming, and can have other compounding factors that can lead to inconsistent evaluation. Hence, human evaluation experiments should be well designed and should be used only when needed and otherwise not. Now, let's look at the three major types of evaluation approaches:

- **Application-grounded evaluation**: This evaluation method involves including the explainability techniques in a real product or application, thus allowing the conduction of human subject experiments, in which real end users are involved to perform certain experiments. Although the experiment setup cost is high and time-consuming, building an almost finished product and then allowing domain experts to test has its benefits. It will enable the researcher to evaluate the quality of the explanation with respect to the end task of the system, thus providing ways to quickly identify errors or limitations of the explainability methods. This evaluation principle is consistent with the evaluation methods used in HCI, and explainability is infused within the entire system responsible for solving the user's keep problem and helping the user meet its end goal.

  For example, to evaluate the quality of explanation methods of an AI software for automated skin cancer detection from images, dermatologists can be approached to directly test the objective for which the AI software is built. If the explanation methods are successful, then such a solution can be scaled up easily. In terms of the industrial perspective, since getting a perfect finished product can be time-consuming, the better approach is to build a robust prototype or a **Minimum Viable Product** (**MVP**) so that the domain expert testing the system gets a better idea of how the finished product will be.

- **Human-grounded evaluation**: This evaluation method involves conducting human subject experiments with non-expert novice users on more straightforward tasks rather than domain experts. Getting domain experts can be time-consuming and expensive, so human-grounded evaluation experiments are easier and less costly to set up. The tasks are also simplified sometimes and usually for certain use cases where generalization is important, these methods are very helpful. **A/B testing, counterfactual simulations**, and **forward simulations** are certain popular evaluation methods used for human-grounded evaluation.

   In XAI, A/B testing provides different types of explanations to the user, where the user is asked to select the best one with the higher quality of explanation. Then, based on the final aggregated votes, and using other metrics such as click-through rate, screen hovering time, and time to task completion, the best method is decided.

   For counterfactual simulation methods, human subjects are presented with the input and output of the model with the model explanations for a certain number of data samples and are asked to provide certain changes to the input features in order to change the model's final outcome to a specific range of values or a specific category. In the forward simulation method, human subjects are provided with the model inputs and their corresponding explanation methods and then asked to simulate the model prediction without looking at the ground-truth values. Then, the error metric used to find the difference between human-predicted outcomes with the ground-truth labels can be used as a quantitative way to evaluate the quality of explanation.

- **Functionality-grounded evaluation**: This evaluation method doesn't involve any human subject experiments, and proxy tasks are used to evaluate the quality of explanation. These experiments are more feasible and less expensive to set up than the other two, and especially for use cases where human subject experiments are restricted and unethical, this is an alternative approach. This approach works well when the type of algorithm was already tested in human-level evaluation.

For example, linear regression models are easily interpretable and end users can efficiently understand the working of the model. So, using a linear regression model for use cases such as sales forecasting can help us to understand the overall trend of the historical data and how the forecasted values are related to the trend.

Figure 1.9 summarizes the taxonomy of the evaluation level for explainable ML systems:

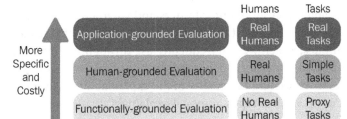

Figure 1.9 – Taxonomy of evaluation level for the explainable ML system

Apart from the methods discussed here, for determining the quality of explanations, other metrics such as description length of explanation, the complexity of the features used in the explanation, and cognitive processing time required to understand the provided explanation, are often used. We will discuss them in more detail in *Chapter 11, End User-Centered Artificial Intelligence.*

In this chapter, so far you have come across many new concepts for ML explainability techniques. The mind-map diagram in *Figure 1.10* gives a nice summary of the various terms and concepts discussed in this chapter:

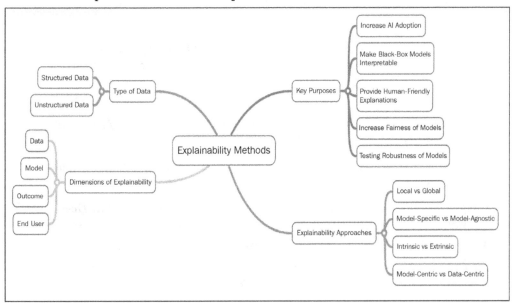

Figure 1.10 – Mind-map diagram of machine learning explainability techniques

I strongly recommend all of you make yourselves familiar with the jargon used in this mind-map diagram as we will be using it throughout the book! Let's summarize what we have discussed in the summary section.

Summary

After reading this chapter, you should now understand what XAI is and why it is so important. You have learned about the various terms and concepts related to explainability techniques, which we will frequently use throughout this book. You have also learned about certain key criteria of human-friendly explainable ML systems and different approaches to evaluating the quality of the explainability techniques. In the next chapter, we will focus on various types of model explainability methods for structured and unstructured data.

References

Please refer to the following resources to gain additional information:

- *The $300m flip flop: how real-estate site Zillow's side hustle went badly wrong, theguardian.com*: https://www.theguardian.com/business/2021/nov/04/zillow-homes-buying-selling-flip-flop

- *Zillow says it's closing homebuying business, cutting 25% of workforce; earnings miss estimates, cnbc.com*: https://www.cnbc.com/2021/11/02/zillow-shares-plunge-after-announcing-it-will-close-home-buying-business.html

- *Google Photos Tags Two African-Americans As Gorillas Through Facial Recognition Software, forbes.com*: https://www.forbes.com/sites/mzhang/2015/07/01/google-photos-tags-two-african-americans-as-gorillas-through-facial-recognition-software/

- *Google apologises for Photos app's racist blunder, bbc.com*: https://www.bbc.com/news/technology-33347866

- *Tutorial on Interpretable Machine Learning, MICCAI'18, by Samek and Binder*: http://www.heatmapping.org/slides/2018_MICCAI.pdf

- *Chapter 1 – Interpretation, Interpretability, and Explainability and Why Does It All Matter?* by *Serg Masis, Packt Publishing Ltd.*: https://www.amazon.com/Interpretable-Machine-Learning-Python-hands/dp/180020390X

- *Interpretable Machine Learning, Christoph Molnar*: https://christophm.github.io/interpretable-ml-book/

- *Towards A Rigorous Science of Interpretable Machine Learning by Doshi-Velez and Kim*: https://arxiv.org/abs/1702.08608

2
Model Explainability Methods

One of the key goals of this book is to empower its readers to design Explainable ML systems that can be used in production to solve critical business problems. For a robust Explainable ML system, explainability can be provided in multiple ways depending on the type of problem and the type of data used. Providing explainability for structured tabular data is relatively human-friendly compared to unstructured data such as images and text, as image or text data is more complex with less interpretable granular features.

There are different ways to add explainability to ML models, for instance, by extracting information about the data or the model (knowledge extraction), using effective visualizations to justify the prediction outcomes (result visualization), identifying dominant features in the training data and analyzing its effect on the model predictions (influence-based methods), or by comparing model outcomes with known scenarios or situations as an example (example-based methods).

So, in this chapter, we are going to discuss various model-agnostic and model-specific explanation methods that are used for both structured and unstructured data for model explainability.

This chapter covers the following main topics:

- Types of model explainability methods
- Knowledge extraction methods
- Result visualization methods
- Influence-based methods
- Example-based methods

Technical requirements

The primary goal of this chapter is to provide a conceptual understanding of the model explainability methods. However, I will provide certain tutorial examples to implement some of these methods in Python on certain interesting datasets. We will be using Python Jupyter notebooks to run the code and visualize the output throughout this book. The code and dataset resources for *Chapter 2* can be downloaded or cloned from the following GitHub repository: `https://github.com/PacktPublishing/Applied-Machine-Learning-Explainability-Techniques/tree/main/Chapter02`. Other important Python frameworks that are required to run the code will be mentioned in the notebooks along with other relevant details to understand the code implementations within these concepts.

Types of model explainability methods

There are different approaches that you can use to provide model explainability. Certain techniques are specific to a model, and certain approaches are applied to the input and output of the model. In this section, we will discuss the different types of methods used to explain ML models:

- **Knowledge extraction methods**: Extracting key insights and statistical information from the data during **Exploratory Data Analysis (EDA)** and post-hoc analysis is one way of providing model-agnostic explainability. Often, statistical profiling methods are applied to extract the mean and median values, standard deviation, or variance across the different data points, and certain descriptive statistics are used to estimate the expected range of outcomes.

Similarly, other insights using correlation heatmaps, decomposition trees, and distribution plots are also used to observe any relationships between the features to explain the model's results. For more complex unstructured data, such as images, often, these statistical knowledge extraction methods are not sufficient. Human-friendly methods using **Concept Activation Vectors** (**CAVs**), as discussed in *Chapter 8, Human-Friendly Explanations with TCAV*, are more effective.

However, primarily, knowledge extraction methods extract essential information about the input data and the output data from which the expected model outcomes are defined. For example, to explain a time series forecasting model, we can consider a model performance metric such as the variance in forecasting error over the training period. The error rate can be specified within a confidence interval of (let's say) +/- 10%. The formation of the confidence interval is only possible after extracting key insights from the output training data.

- **Result visualization methods**: Plotting model outcomes and comparing them with previously predicted values, particularly with surrogate models, is often considered an effective model-agnostic explainability method. Predictions from black-box ML algorithms are passed to the surrogate model explainers. Usually, these are highly interpretable linear models, decision trees, or any rule-based heuristic algorithm that can explain the outcome of complex models. The main limitation of this approach is that the explainability is solely dependent on the model outcome. If there is any abnormality with the data or the modeling process, such dimensions of explainability are not captured.

  For example, let's suppose a classifier is incorrectly predicting an output. It is not feasible for us to understand exactly why the model is behaving in a specific manner just from the prediction probability. But these methods are easy to apply in practice and even easy to understand as highly interpretable explainer algorithms are used.

- **Influence-based methods**: These are specific techniques that help us to understand how certain data features play an important role in influencing or controlling the model outcome. Right now, this is one of the most common and effective methods applied to provide explainability to ML models. Feature importance, sensitivity analysis, key influencer maps, saliency maps, **Class Activation Maps** (**CAMs**), and other visual feature maps are used to interpret how the individual features within the data are being utilized by the model for its decision-making process.

- **Example-based methods**: The three previously discussed model-agnostic explainability methods still need some kind of technical knowledge to understand the working of the ML models. For non-technical users, the best way to explain something is to provide an example that they can relate to. Example-based methods, particularly counterfactual example-based methods, try to look at certain single instances of the data to explain the ML models' decision-making process.

  For example, let's say an automated loan approval system powered by ML denies a loan request to an applicant. Using example-based explainability methods, the applicant will also be suggested that if they pay their credit card bill on time for the next three months and increase their monthly income by $2,000, their loan request would be granted.

Based on the latest trends, model-agnostic techniques are preferred over model-dependent approaches as even complex ML algorithms can be explained, to some degree, using these techniques. But certain techniques such as saliency maps, tree/forest-based feature importance, and activation maps are mostly model-specific. Our choice of explanation method is determined by the key problem that we are trying to solve.

Figure 2.1 illustrates the four main types of explainability methods that have been applied to explain the working of black-box models, which we are going to cover in the following sections:

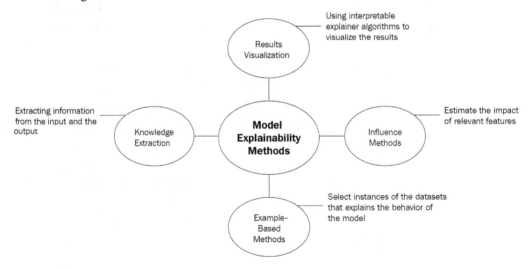

Figure 2.1 – Model explainability methods

Now, let's start by discussing each of these model explainability methods in more detail.

Knowledge extraction methods

Whenever we talk about explainability in any context, it is all about gaining knowledge of the problem so as to gain some clarity about the expected outcome. Similarly, if we already know the outcome, explainability is all about tracing back to the root cause. Knowledge extraction methods in ML are used to extract key insights from the input data or utilize the model outcome to trace back and map to certain information known to the end users for both structured data and unstructured data. Although there are multiple approaches to extracting knowledge, in practice, the data-centric process of EDA is one of the most common and popular methods for explaining any black-box model. Let's discuss more on how to use the EDA process in the context of **XAI**.

EDA

I would always argue that EDA is the most important process for any ML workflow. EDA allows us to explore the data and draw key insights; using this, we can form certain hypotheses from the data. This actually helps us to identify any distinct patterns within the data and will, eventually, help us to make the right choice of algorithm. Thus, EDA is one of the conventional and model-agnostic approaches that explain the nature of the data, and by considering the data, it helps us understand what to expect from the model. Detecting any clear anomaly, ambiguous, redundant data points and bias in data can be easily observed using EDA. Now, let's see some important methods used in EDA to explain ML models for structured and unstructured data.

EDA on structured data

EDA on structured data is one of the preliminary steps applied for extracting insights to provide explainability. However, the actual techniques applied in the EDA process could vary from one problem to another. But generally, for structured data, we can use EDA to generate certain descriptive statistics for a better understanding of the data and then apply various univariate and multivariate methods to detect the importance of each feature, observe the distribution of data to find any biases in the data, and look for outliers, duplicate values, missing values, correlation, and cardinality between the features, which might impact the model's results.

Information and hypotheses obtained from the EDA step help perform meaningful feature engineering and modeling techniques and help set up the right expectation for the stakeholders. In this section, we will cover the most popular EDA methods and discuss the benefit of using EDA with structured data in the context of XAI. I strongly recommend looking at the GitHub repository (`https://github.com/PacktPublishing/Applied-Machine-Learning-Explainability-Techniques`) to apply some of these techniques in practice for practical use cases. Now, let's look at the important EDA methods in the following list:

- **Summary statistics**: Usually, model explainability is presented with respect to the features in the data. Observing dataset statistics during the EDA process gives an early indication of whether the dataset is sufficient for modeling and solving the given problem. It helps to understand the dimensions of the data and the type of features present. If the features are numeric, certain descriptive statistics such as the mean, standard deviation, coefficient of variation, skewness, kurtosis, and inter-quartile ranges are observed.

  Additionally, certain histogram-based distributions are used to monitor any skewness or biases in data. For categorical features, the frequency distribution of the categorical values is observed. If the dataset is imbalanced, if the dataset is biased toward a particular categorical value, if the dataset has outliers, or is skewed toward a particular direction, all of these can be easily observed. Since all of these factors can impact the model predictions, understanding dataset statistics is important for model explainability.

  Figure 2.2 shows the summary statistics and visualizations created during the EDA step to extract knowledge about the data:

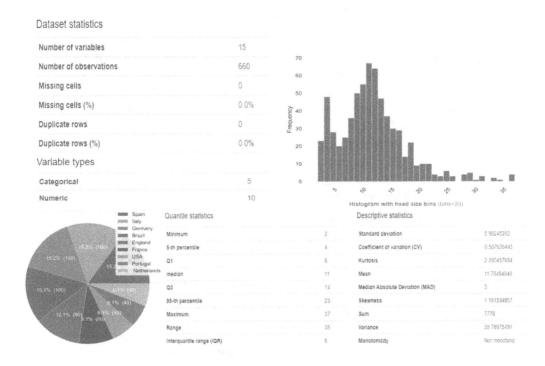

Figure 2.2 – Summary statistics and visualizations during EDA

- **Duplicate and missing values**: Duplicate or redundant values can add more bias to the model. In contrast, missing values can lead to a loss of information and insufficient data to train the model. This might lead to model-overfitting. So, before training the model, if missing values or duplicate values are observed, and if further actions are not taken to rectify this, then these observations might help to explain the reason behind the non-generalization of models.

- **Univariate analysis**: This involves analyzing a single feature through graphical techniques such as distribution plots, histograms, box plots, violin plots, pie charts, clustering plots, and using non-graphical techniques such as frequency, central tendency measures (that is, the mean, standard deviation, and coefficient of variation), and interquartile ranges. These methods help us to estimate the impact of individual features on the model outcome.

- **Multivariate analysis**: This involves analyzing two or more features together using graphical and non-graphical methods. It is used for identifying data correlation and the dependencies of variables. In the context of XAI, multivariate analysis is used to understand complex relationships in the data and provide a detailed and granular explanation as compared to univariate analysis methods.

- **Outlier detection**: Outliers are certain abnormal data points that can completely skew the model. It is hard to achieve generalization if a model is trained on outlier data points. However, model prediction can go completely wrong during the model inference time for an anomaly datapoint. Hence, outlier detection during both training and inference time is an important part of model explainability. Visualization methods such as box plots, scatter plots, and statistical methods such as the 1.5xIQR rule (`https://www.khanacademy.org/math/statistics-probability/summarizing-quantitative-data/box-whisker-plots/a/identifying-outliers-iqr-rule`) and Nelson's Rule (`https://www.leansixsigmadefinition.com/glossary/nelson-rules/`) are used for detecting anomalies.

- **Pareto analysis**: According to the Pareto Principle, 80% of the value or impact is driven by 20% of the sample size. So, in XAI, this *80–20 rule* is used to interpret the most impactful sub-samples that have the maximum impact on the model outcome.

- **Frequent Itemset Mining**: This is another popular choice of approach to extract model explainability. This technique is used frequently for Association rule mining to understand how frequently certain observations occur together in any given dataset. This method provides some interesting observations that help to form important hypotheses from the data and, eventually, contribute a lot to explaining model outcomes.

Now that we have covered the methods for structured data, let's take a look at some of the methods for unstructured data.

EDA on unstructured data

Interpreting features from unstructured data such as images and text is difficult, as ML algorithms try to identify granular-level features that are not intuitively explainable to human beings. Yet, there are certain specific methods applied to image and text data to form meaningful hypotheses from the data. As discussed earlier, the EDA process might change based on the problem and the data, but in this chapter, we will discuss the most popular choice of methods in the context of XAI.

Exploring image data

EDA methods used for images are different from the methods used with tabular data. These are some popular choices of EDA steps for image datasets:

- **Data dimension analysis**: For consistent and generalized models, understanding data dimension is important. Monitoring the number of images and the shape of each image is important to explain any observation of overfitting or underfitting.

- **Observing data distribution**: Since the majority of problems solved using images are classification problems, monitoring class imbalance is important. If the distribution of data is not balanced, then the model can be biased toward the majority class. For pixel-level classification (for segmentation problems), observing pixel intensity distribution is important. This also helps in understanding the effect of shadow or non-uniform lighting conditions on images.

- **Observing average images and contrast images**: For observing dominant regions of interest in images, average and contrast images are often used. This is especially used for classification-based problems to compare dominant regions of interest.

- **Advanced statistical and algebraic methods**: Apart from the methods discussed so far, other statistical methods such as finding the z-score and standard deviation, and algebraic methods such as Eigenimages based on eigenvectors are used to visually inspect key features in image data, which adds explainability to the final model outcome.

There are other complex methods to explore image datasets depending on the type of the problem. However, the methods discussed in this subsection are the most common approaches.

Exploring text data

Usually, text data is noisier in comparison to images or tabular datasets. Hence, EDA is usually accompanied by some preprocessing or cleaning methods for text data. But since we are focusing only on the EDA part, the following list details some of the popular approaches to do EDA with text data:

- **Data dimension analysis**: Similar to images, text dimension analyses, such as checking the number of records and the length of each record, are performed to form hypotheses about potential overfitting or underfitting.

- **Observing data distribution**: Visualizing the distribution of word frequency using bar plots or word clouds are popular choices in which to observe top words in any text data. This technique allows us to avoid any bias of high-frequency words as compared to low-frequency words.

- **n-gram analysis**: Considering the nature of text data, often, a phrase or collection of words is more interpretable than only a single word. For example, for sentiment analysis from movie reviews, individual words with a high frequency such as *movie* or *film* are quite ambiguous. In contrast, phrases such as *good movie* and *very boring film* are far more interpretable and useful to understand the sentiments. Hence, n-gram analysis or taking a collection of "n-words" brings more explainability for understanding the model outcome.

Usually, EDA does include certain visualization techniques to explain and form some important hypotheses from the data. But another important technique to explain ML models is by visualizing the model outcome. In the next section, we will discuss these result visualization methods in more detail.

Result visualization methods

Visualization of the model outcomes is a very common approach applied to interpret ML models. Generally, these are model-agnostic, post-hoc analysis methods applied on trained black-box models and provide explainability. In the following section, we will discuss some of the commonly used result visualization methods for explaining ML models.

Using comparison analysis

These are mostly post-hoc analysis methods that are used to add model explainability by visualizing the model's predicted output after the training process. Mostly, these are model-agnostic approaches that can be applied to both intrinsically interpretable models and black-box models. Comparison analysis can be used to produce both global and local explanations. It is mainly used to compare different possibilities of outcomes using various visualization methods.

For example, for classification-based problems, certain methods such as t-SNE and PCA are used to visualize and compare the transformed feature spaces of the model predicted labels, especially when the error rate is high. For regression and time series predictive models, confidence levels are used to compare model predicted results with the upper and lower bounds. There are various methods to apply comparison analysis and get a clearer idea of the *what-if* scenarios. Some prominent methods are mentioned in the project repository (`https://github.com/PacktPublishing/Applied-Machine-Learning-Explainability-Techniques/blob/main/Chapter02/Comparison%20Analysis.ipynb`).

As we can see in *Figure 2.3*, result visualization can help to provide a global perspective about the model to visualize model predictions:

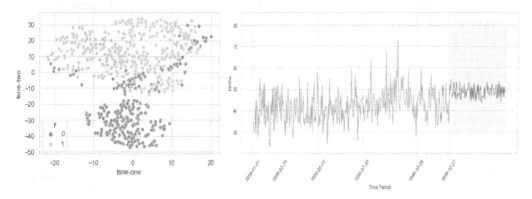

Figure 2.3 – Comparison analysis using the t-SNE method for the classification problem (left-hand side) and the time series prediction model with the confidence interval (right-hand side)

In *Figure 2.3*, we can see how visualization methods can be used to provide local explanations by visualizing the final outcome of the model and comparing the outcome with either other data instances or with possible *what-if* scenarios to provide a global perspective of the model.

Using Surrogate Explainer methods

In the context of ML, when an external model or algorithm is applied to interpret a black-box ML model, the external method is known as the **Surrogate Explainer method**. The fundamental idea behind this approach is to apply an intrinsically explainable model that is simple and easy to interpret and approximate the predictions of the black-box model as accurately as possible. Then, certain visualization techniques are used to visualize the outcome from the Surrogate Explainer methods to get insights into the model behavior.

But now the question is *can we apply surrogate models directly instead of using the black-box model?* The answer is *no!* The main idea behind using the surrogate model is to get some information about how the input data is related to the target outcomes, without considering the model accuracy. In contrast, the original black-box model is more accurate and efficient but not interpretable. So, replacing the black-box model completely with the surrogate model would compromise the model accuracy, which we don't want.

Interpretable algorithms such as regression, decision trees, and rule-based algorithms are popular choices for Surrogate Explainer methods. To provide explainability, mainly three types of relationships between the input features and the target outcome are analyzed: linearity, monotonicity, and interaction.

Linearity helps us to inspect whether the input features are linearly related to the target outcome. Monotonicity helps us to analyze whether increasing the overall input feature values leads to either an increase or a decrease in the target outcome. For the entire range of features, this explains whether the relationship between the input features and the target always propagates in the same direction. Model interactions are extremely helpful when providing explainability, but it is tough to achieve. Interactions help us to analyze how individual features interact with each other to impact the model decision-making process.

Decision trees and rule-based algorithms are used to inspect interactions between the input features and the target outcome. *Christoph Molner*, in his book *Interpretable Machine Learning* (https://christophm.github.io/interpretable-ml-book/), has provided a very useful table comparing different intrinsically interpretable models, which can be useful for selecting interpretable models as a Surrogate Explainer model.

A simplified version of this is shown in the following table:

Algorithm	Relationship	Type of ML problem
Decision trees	Interactions and monotonicity (to some extent)	Classification and regression
RuleFit	Linearity and interactions	Classification and regression
Linear regression	Linearity and monotonicity	Regression
Logistic regression	Monotonicity	Classification
Naïve Bayes	Monotonicity	Classification

Figure 2.4 – Comparing interpretable algorithms for selecting a Surrogate Explainer method

One of the major advantages of this technique is that it can help to make any black-box model interpretable. It is model-agnostic and very easy to implement. But when the data is complex, more sophisticated algorithms are used to achieve higher modeling accuracy. In such cases, surrogate methods tend to oversimplify complicated patterns or relationships between the input features.

Figure 2.5 illustrates how interpretable algorithms such as decision trees, linear regression, or any heuristic rule-based algorithm can be used as surrogate models to explain any black-box ML model:

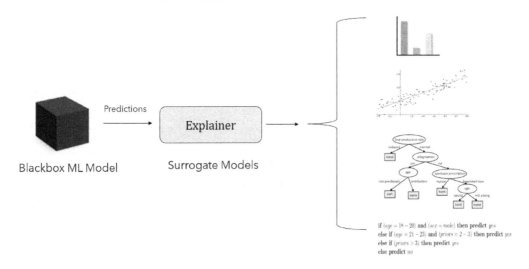

Figure 2.5 – Using Surrogate explainers for model explainability

Despite the drawbacks of this approach, visualization of linearity, monotonicity, and interaction can justify working on complex black-box models to a great extent. In the next section, we will discuss influence-based methods to explain ML models.

Influence-based methods

Influence-based methods are used to understand the impact of features present in the dataset on the model's decision-making process. Influence-based methods are widely used and preferred in comparison to other methods as this helps to identify the dominating attributes from the dataset. Identifying the dominating attributes from structured and unstructured data helps us analyze the dominating features' role in influencing the model outcome.

For example, let's say you are working on a classification problem for classifying wolves and Siberian huskies. Let's suppose that after the training and evaluation process, you have achieved a good model with more than 95% accuracy. But when trying to find the important features using influence-based methods for model explainability, you observed that the model picked up the surrounding background as the dominating feature to classify whether it is a wolf or a husky. In such cases, even if your model result seems to be highly accurate, your model is unreliable. This is because the features that the model was making the decision on are not robust and generalized.

Influence-based methods are popularly used for performing root cause analysis to debug ML models and detect failures in ML systems. Now, let's discuss some of the popular choices of the influence-based methods that are used for model explainability.

Feature importance

When applying an ML model, understanding the relative importance of each feature in terms of influencing the model outcome is crucial. It is a technique that assigns a particular score to the input features present in the dataset based on the usefulness of the features in predicting the target value. Feature importance is a very popular choice of model-agnostic explainability for modeling structured data. Although there are various scoring mechanisms to determine feature importance, such as permutation importance scores, statistical correlation scores, decision tree-based scoring, and more, in this section, we will mostly focus on the overall method and not just on the scoring mechanism.

In the context of XAI, feature importance can provide global insights into the data and the model behavior. It is often used for feature selection and dimensionality reduction to improve the efficiency of ML models. By removing less important features from the modeling process, it has been observed that, usually, the overall model performance is improved.

The notion of important features can sometimes depend on the type of scoring mechanism or the type of model used. So, it is recommended that you validate the important features picked up by this technique with domain experts before drawing any conclusions. This method is applied to structured datasets, where the features are clearly defined. For unstructured data such as text or images, feature importance is not very relevant as the features or patterns used by the model are more complex and not always human interpretable.

Figure 2.6 illustrates how highlighting the influential features of a dataset enables the end user to focus on the values of the important features to justify the model outcome:

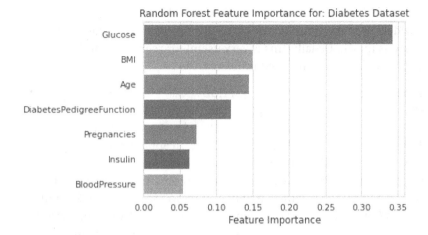

Figure 2.6 – A feature importance graph on the diabetes dataset (from the code tutorials)

Next, we will cover another important influence-based model explainability method, called sensitivity analysis.

Sensitivity analysis

Sensitivity analysis is a quantitative process that approximates uncertainty in forecasts by altering the assumptions made about important input features used by the forecasting model. In sensitivity analysis, the individual input feature variables are increased or decreased to assess the impact of the individual features on the target outcome. This technique is very commonly used in predictive modeling to optimize the overall performance and robustness of the system.

Conducting sensitivity analysis can be simple yet a very powerful method for any data science project, which can provide additional information to business stakeholders, especially for multivariate datasets. It helps to understand the *what-if* scenarios and observe whether any particular feature is sensitive to outliers or any form of adversarial perturbations. It helps in questioning the reliability of variable assumptions, can predict the possible outcomes if the assumptions are changed, and can measure the significance of altering variable assumptions. Sensitivity analysis is a data-driven modeling approach. It indicates whether the data is reliable, accurate, and relevant for the modeling process. Additionally, it helps to find out whether there are other intervening factors that can impact the model.

In the context of XAI, since sensitivity analysis is slightly less common as compared to some of the widely used methods, let me try to give my recommendations for performing sensitivity analysis in ML. Usually, this is very useful for regression problems, but it is quite important for classification-based problems, too.

Please refer to the notebook (`https://github.com/PacktPublishing/Applied-Machine-Learning-Explainability-Techniques/blob/main/Chapter02/FeatureImportance_SensitivityAnalysis.ipynb`) provided in the GitHub repository to get a detailed and practical approach for doing sensitivity analysis. The very first step that I recommend you do for sensitivity analysis is to calculate the standard deviation (σ) of each attribute that is present in the raw dataset. Then, for each attribute, transform the original attribute values to -3σ, -2σ, -σ, σ, 2σ, and 3σ, and either observe and plot the percentage change in the target outcome for a regression problem or observe the predicted class for a classification-based problem.

For a good and robust model, we would want the target outcome to be less sensitive to any change in the feature values. Ideally, we would expect the percentage change in the target outcome to be less drastic for regression problems, and for classification problems, the predicted class should not change much on changing the feature values. Any feature value beyond +/- 3σ is considered an outlier, so usually, we vary the feature values up to +/- 3σ.

Figure 2.7 shows how detailed sensitivity analysis helps you analyze factors that can easily influence the model outcome:

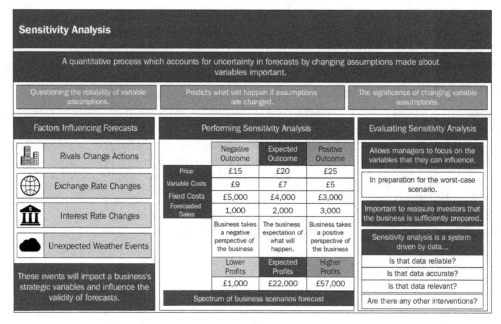

Figure 2.7 – Sensitivity analysis to understand the influential features of the data

Apart from sensitivity analysis, in the next section, you will learn about **Partial Dependence Plots** (**PDPs**), which can also be used to analyze influential features.

PDPs

When using black-box ML models, inspecting functional relations between feature attributes and target outcomes can be challenging. Although calculating feature importance can be easier, PDPs provide a mechanism to functionally calculate the relationship between the predictive features and the predictor variables. It shows the marginal effect one or two attributes have on the target outcome.

PDPs can effectively help pick up linear, monotonic, or any complex interaction between the predictive variables and the predictor variables and indicate the overall impact that the predictor variable has on the predictive variable on average. PDP includes the contribution of particular predictor attributes by measuring the marginal effect, which does not include the other variables' impact on the feature space.

Similar to sensitivity analysis, PDPs help us to approximate the direction in which specific features can influence the target outcome. For the sake of simplicity, I will not add any complex mathematical representation of partial dependence to obtain the average marginal effect of predictor variables; however, I would strongly recommend going through *Jerome H. Friedman*'s work on *Greedy Function Approximation: A Gradient Boosting Machine* to get more information. One of the most significant benefits of PDP is that it is straightforward to use, implement, and understand and can be explained to a non-technical business stakeholder or an end user of ML models very easily.

But there are certain drawbacks to this approach, too. By default, the approach assumes that all features are not correlated and there is no interaction between the feature attributes. In any practical scenario, this is highly unlikely to happen, as the majority of the time, there will be some interaction or joint effect due to the feature variables.

PDPs are also limited to two-dimensional representations, and PDPs do not show any feature distribution. So, if the feature space is not evenly distributed, certain effects of bias can get missed while analyzing the outcome. PDPs might not show any heterogeneous effects as it only shows the average marginal effects. This means that if half of the data for a particular feature has a positive impact on the predicted outcome, and the other half has a negative effect on the predicted outcome, then the PDP could just be a horizontal line as the effects from both halves can cancel each other. This can lead to the conclusion that the feature does not have any impact on the target variable, which is misleading.

The drawbacks of PDP can be solved by **Accumulated Local Effect Plots** (**ALEP**) and **Individual Conditional Expectation Curves** (**ICE curves**). We will not be covering these concepts in this chapter, but please refer to the *Reference* section, *[Reference – 4,5]*, to get additional resources to help you understand these concepts.

Let's look at some sample PDP visualizations from *Figure 2.8*:

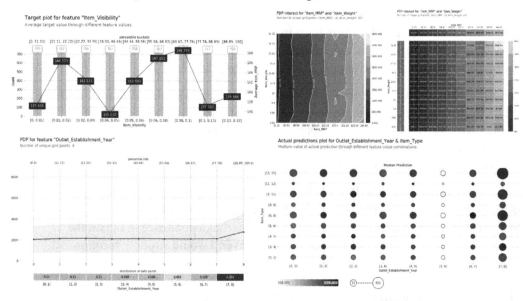

Figure 2.8 – PDP visualizations (from the code tutorial)

Figure 2.8 illustrates PDP visualizations that help us to understand influential features from tabular datasets. In the next section, we will discuss the **Layer-wise Relevance Propagation** (**LRP**) methods to understand influential features from unstructured data.

LRP

Most of the influence-based methods that we discussed earlier are highly effective for structured data. But unfortunately, these methods cannot be applied to unstructured data such as images and texts where the features are not always clearly defined, especially when using **Deep Convolution Neural Networks** (**DCNNs**). Classical ML algorithms are not efficient as compared to deep learning algorithms when applied to unstructured data such as images and text. Due to the benefit of automatic feature extraction in deep learning as compared to manual feature engineering in classical ML, deep learning algorithms are more efficient in terms of model accuracy and, hence, more preferred. However, deep learning models are more complex and less interpretable than classical ML models.

Providing explainability to deep learning models is also quite challenging; usually, there are very few quantitative ways for providing explainability to deep learning models. Therefore, we mostly rely on qualitative approaches to visualize the key influencing data elements that can impact the process of calculating weights and biases, which are the main parameters of any deep learning model. Moreover, for deep networks with multiple layers, learning happens when the flow of information through the gradient flow process between the layers is maintained consistently. So, to explain any deep learning model, particularly for images and text, we would try to visualize the *activated* or most influential data elements throughout the different layers of the network and qualitatively inspect the functioning of the algorithm.

To explain deep learning models, LRP is one of the most prominent approaches. Intuitively speaking, this method utilizes the weights in the network and the forward pass neural activations to propagate the output back to the input layer through the various layers in the network. So, with the help of the network weights, we can visualize the data elements (pixels in the case of images and words in the case of text data) that contributed most toward the final model output. The contribution of these data elements is a qualitative measure of relevance that gets propagated throughout the network layers.

Now, we will explore some specific LRP methods that have been applied to explain the working of deep learning models. In practice, implementing these methods can be challenging. So, I have not included these methods in the code tutorials, as this chapter is supposed to help even beginner learners. I have shared some resources in the *Reference* section for intermediate or advanced learners for the code walk-throughs.

Saliency maps

A saliency map is one of the most popularly used approaches for interpreting the prediction of **Convolution Neural Networks (CNNs)**. This technique is derived from the concept of image saliency, which refers to the important features of an image, such as the pixels, which are visually alluring. So, a saliency map is another image derived from the original image in which the pixel brightness is directly proportional to the saliency of the image. A saliency map helps to highlight regions within the image that play an important role in the final decision-making process for the model. It is a visualization technique used specifically in DCNN models to differentiate visual features from the data.

Apart from providing explainability to deep learning models, saliency maps can be used to identify regions of interest, which can be further used by automated image annotation algorithms. Also, saliency maps are used in the audio domain, particularly in audio surveillance, to detect unusual sound patterns such as gunshots or explosions.

Figure 2.9 shows the saliency map for a given input image, which highlights the important pixels used by the model to predict the outcome:

Figure 2.9 – Saliency maps for an input image

Next, let's cover another popular LRP method – **Guided Backpropagation (Guided Backprop)**.

Guided backprop

Another visualization technique used for explaining deep learning models to increase trust and their adoption is **guided backprop**. Guided backprop highlights granular visual details in an image to interpret why a particular class was predicted by the model. It is also known as guided saliency and actually combines the process of vanilla backpropagation and backpropagation through ReLU nonlinearity (also referred to as **DeconvNets**). I would strongly recommend looking at this article, `https://towardsdatascience.com/review-deconvnet-unpooling-layer-semantic-segmentation-55cf8a6e380e,` to learn more about the backpropagation and the DeconvNet mechanism if you are not aware of these terms.

In this method, the neurons of the network act as the feature detectors and, because of the usage of the ReLU activation function, only gradient elements that are positive in the feature map are kept. Additionally, the DeconvNets only keep the positive error signals. Since the negative gradients are set to zero, only the important pixels are highlighted when backpropagating through the ReLU layers. Therefore, this method helps to visualize the key regions of the image, the vital shapes, and the contours of the object that are to be classified by the algorithm.

Figure 2.10 shows a guided backprop map for a given input image that marks the contours and some granular visual features used by the model to predict the outcome:

Border terrier   **Guided Backprop**

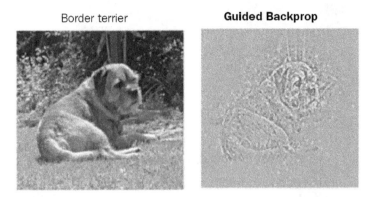

Figure 2.10 – Guided backprop for an input image

Guided backprop is very useful, but in the next section, we will cover another useful method to explain unstructured data such as images, called the Gradient CAM.

Gradient CAM

CAMs are separate visualization methods used for explaining deep learning models. Here, the model predicted class scores are traced back to the last convolution layer to highlight discriminative regions of interest in the image that are class-specific and not even generic to other computer vision or image processing algorithms. **Gradient CAM** combines the effect of guided backprop and CAM to highlight class discriminative regions of interest without highlighting the granular pixel importance, unlike guided backprop. But Grad-CAM can be applied to any CNN architectures, unlike CAM, which can be applied to architectures that perform global average pooling over output feature maps coming from the convolution layer, just prior to the prediction layer.

Grad-CAM (also referred to as **Gradient-Weighted Class Activation Map**) helps to visualize high-resolution details, which are often superimposed on the original image to highlight dominating image regions for predicting a particular class. It is extremely useful for multi-class classification models. Grad-CAM works by inspecting the gradient information flow into the last layer of the model. However, for certain cases, it is important to inspect fine-grained pixel activation information, too. Since Grad-CAM doesn't allow us to inspect granular information, there is another variant of Grad-CAM, which is known as **Guided Grad-CAM**, used to combine the benefits of guided backprop with Grad-CAM to even visualize the granular-level class discriminative information in the image.

Figure 2.11 shows what a Grad-CAM visualization looks like for any input image:

Figure 2.11 – Grad-CAM for an input image

Grad-CAM highlights the important regions in the image, which are used by the model to predict the outcome. Another variant of this approach is to use guided Grad-CAM, which combines the guided backpropagation and Grad-CAM methods to produce interesting visualizations to explain deep learning models:

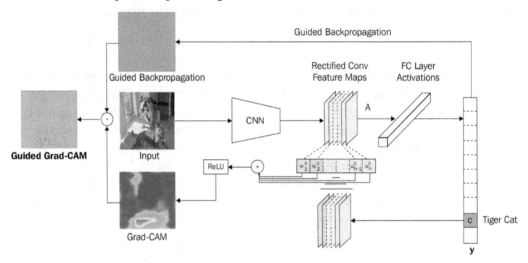

Figure 2.12 – Architecture diagram for guided Grad-CAM

Figure 2.12 shows the architecture diagram for the guided Grad-CAM approach that is slightly more complex to understand. But overall, LRP is an important approach that can be used to explain the functioning of deep learning models.

Representation-based explanation

Pattern representation plays an important role in the decision-making process, especially for unstructured data such as text and images. Conventionally, hand-engineered pattern matching algorithms were used for extracting global features, which human beings can relate to. But recently, GoogleAI's model interpretability technique, which is based on CAVs, gained great popularity in the field of XAI. In this part, we will discuss CAVs in more detail, although extracting features and patterns from unstructured data also falls under the representation-based explanations.

CAVs

Particularly for unstructured data, most deep learning models work on low-level features such as edges, contours, and motifs, and some mid-level and high-level features such as certain defined parts and portions of the object of interest. Most of the time, these representations are not human-friendly, especially for complex deep learning models. Intuitively, CAVs relate the presence of low-level and granular features to high-level human-friendly concepts. Therefore, model explainability with CAVs provides more realistic explanations to which any human being can relate.

The approach of CAVs is actually implemented using the **Testing with Concept Activation Vectors** (**TCAV**) framework from GoogleAI. TCAV utilizes directional derivatives to approximate the internal state of the neural network to a human-defined concept. For example, if we ask a human being to explain what a zebra looks like, they would probably say that a zebra is an animal that looks like a white horse with black stripes and is found in grasslands. So, the terms *animal*, *white horse*, *black stripes*, and *grasslands* can be important concepts used to represent zebras.

Similarly, the TCAV algorithm tries to learn these concepts and learn how much of the concept was important for the prediction using the trained model, although these concepts might not be used during the training process. So, TCAV tries to quantify how sensitive the model is toward the particular concept for a particular class. I found the idea of CAVs very appealing. I think it is a step toward creating human-friendly explanations of AI models, which any non-technical user can easily relate to. We will be discussing the TCAV framework from GoogleAI, in more detail, in *Chapter 8, Human-Friendly Explanations with TCAV.*

Figure 2.13 illustrates the idea of using human-friendly concepts to explain model predictions. In the next subsection, we will see another visualization approach that is used to explain complex deep learning models:

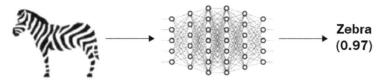

How important is the notion of "stripes" for this prediction?

Figure 2.13 – The fundamental idea behind the CAV

Next, we will discuss **Visual Attention Maps** (**VAMs**), which can also be used with complicated deep learning models.

VAMs

In recent years, transformer model architectures have gained a lot of popularity because of their ability to achieve state-of-the-art model performance on complicated unstructured data. Attention networks are the heart of transformer architecture, which allows the algorithm to learn more contextual information for producing more accurate outcomes.

The basic idea is that every portion of the data is not equally important and only the important features need more *attention* than the rest of the data. Therefore, the attention network filters out irrelevant portions of the data for making a better judgment. By the attention mechanism, the network can assign higher weights to the important sections by the level of importance to the underlying task. Using these attention weights, certain visualizations can be created which explain the decision-making process of the complex algorithm. These are called VAMs.

This technique is particularly useful for **Multimodal Encoder-Decoder** architectures for solving problems such as automated image captioning and visual question answering. Applying VAMs can be quite complicated if you have a beginner level of understanding. So, I will not cover this technique in much detail in this book or in the code tutorials. If you are interested in learning more about how this technique work in practice, please refer to the code repository at `https://github.com/sgrvinod/a-PyTorch-Tutorial-to-Image-Captioning`.

As we can see in *Figure 2.14*, VAMs provide step-by-step visuals to explain the output of complex encoder-decoder models. In the next section, we will explore example-based methods, which are used to explain ML models:

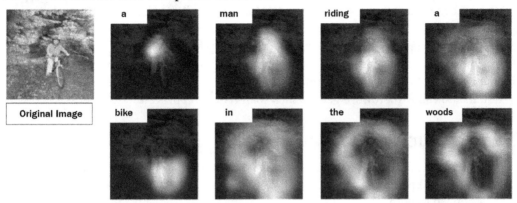

Figure 2.14 – Using VAMs to explain complex encoder-decoder attention-based deep learning models for the task of automated image captioning using a multi-modal dataset

In the next section, we will cover another type of explainability method that uses human-friendly examples to interpret predictions from black-box models.

Example-based methods

Another approach to model explainability is provided by example-based methods. The idea of example-based methods is similar to how humans try to explain a new concept. As human beings, when we try to explain or introduce something new to someone else, often, we try to make use of examples that our audience can relate to. Similarly, example-based methods, in the context of XAI, try to select certain instances of the dataset to explain the behavior of the model. It assumes that observing similarities between the current instance of the data with a historic observation can be used to explain black-box models.

These are mostly model-agnostic approaches that can be applied to both structured and unstructured data. If the structured data is high-dimensional, it becomes slightly challenging for these approaches, and all the features cannot be included to explain the model. So, it works well only if there is an option to summarize the data instance or pick up only selected features.

In this chapter, we will mainly discuss **Counterfactual Explanations (CFEs)**, the most popular example-based explainability method that works for both structured and unstructured data. CFEs indicate to what extent a particular feature has to change to significantly change the predicted outcome. Typically, this is useful for classification-based problems.

For certain predictive models, CFEs can provide prescriptive insights and recommendations that can be very crucial for end users and business stakeholders. For example, let's suppose there is an ML model used in an automated loan approval system. If the black-box model denies the loan request for a particular applicant, the loan applicant might reach out to the provider to learn the exact reason why their request was not approved. But instead, if the system suggests the applicant increases their salary by 5,000 and pay their credit card bills on time for the next 3 months in order to approve the loan request, then the applicant will understand and trust the decision-making process of the system and can work toward getting their loan request approved.

CFEs in structured data

Using the **Diverse Counterfactual Explanation (DiCE)** framework in Python (`https://interpret.ml/DiCE/`), a CFE can be provided for structured data. It can be applied to both classification and regression-based problems for model-agnostic local explainability, and it describes how the smallest change in the structured data can change the target outcome.

Another important framework used for CFEs in Python is **Alibi** (`https://docs.seldon.io/projects/alibi/en/stable/`), which is also pretty good in terms of implementing the concepts of CFE to explain ML models. Although we will be discussing these frameworks and experiencing the practical aspects of these frameworks in *Chapter 9, Other Popular XAI Frameworks*, for now, I will discuss some intuitive understanding of CFE on structured data. Please refer to the notebook (`https://github.com/PacktPublishing/Applied-Machine-Learning-Explainability-Techniques/blob/main/Chapter02/Counterfactual_structured_data.ipynb`) provided in the code repository to learn how to code these approaches in Python for practical problem-solving.

When used with structured data, the CFE method tries to analyze an input query data instance and tries to observe the original target outcome considering the same query instance from the historical data. Alternatively, it tries to inspect the features and maps them to a similar instance present in the historical data to get the target output. Then, the algorithm generates multiple counterfactual examples to predict the opposite outcome.

For a classification-based problem, this method would try to predict the opposite class for a binary classification problem or the nearest or most similar class for a multiclass classification problem. For a regression problem, if the target outcome is present toward the lower end of the spectrum, the algorithm tries to provide a counterfactual example with a target outcome closer to the higher end of the spectrum, and vice versa. Hence, this method is very effective for understanding *what-if* scenarios and can provide actionable insights along with model explainability. The explanations are also very clear and very easy to interpret and implement.

But the major drawback of this approach, particularly for structured data, is that it suffers from the *Rashomon effect* (`https://www.dictionary.com/e/pop-culture/the-rashomon-effect/`). For any real-world problem, it can find multiple CFEs that can contradict each other. With structured data, with multiple features, contradictory CFEs can create more confusion rather than explaining ML models! Human intervention and the application of domain knowledge to pick up the most relevant example can help in mitigating the Rashomon effect. Otherwise, my recommendation is to combine this method along with the feature importance method for actionable features to select counterfactual examples involving significant changes for providing better actionable explainability.

Figure 2.15 illustrates how CFEs can be used to get prescriptive insights and actionable recommendations to explain the working of models:

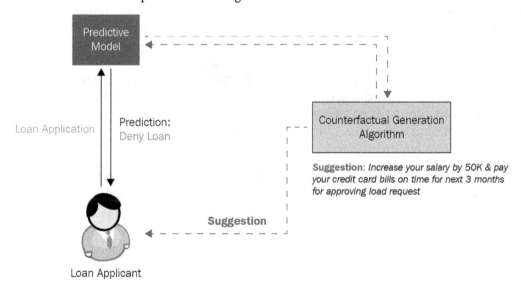

Figure 2.15 – Prescriptive insights obtained from CFEs in structured data

CFEs in tabular datasets can be very useful as they can provide actionable suggestions to the end user. In the next subsection, we will explore CFEs in unstructured data.

CFEs in unstructured data

In unstructured data such as images and text, implementing CFEs can be quite challenging. One of the main reasons for this is that the granular features used in images or text by deep learning models are not always well-defined or human-friendly. But the **Alibi** (https://docs.seldon.io/projects/alibi/en/stable/) framework does pretty well in generating CFEs on image data. Even the improved version of the simple CFE method performs even better by generating a CFE guided by class prototypes. It uses an auto-encoder or k-d trees to build a prototype for each prediction class using a certain input instance.

For example, in the MNIST dataset, let's suppose that the input query image is for digit 7. Then, the counterfactual prototype method will build prototypes of all the digits from 0 to 9. Following this, it will try to produce the nearest digit other than the original digit of 7 as the counterfactual example. Depending upon the data, the nearest hand-written digit can be either 9 or 1; even as human beings, we might confuse the digits 7 and 9 or 7 and 1 if the handwriting is not clear! It takes an optimization approach to minimize the model's counterfactual prediction loss.

I strongly recommend looking at *Arnaud Van Looveren* and *Janis Klaise's* work, *Interpretable Counterfactual Explanations Guided by Prototypes* (https://arxiv.org/abs/1907.02584), to get more details on how this approach works. This CFE method, guided by prototypes, also eliminates any computational constraint that can arise due to the numerical gradient evaluation process for black-box deep learning models. Please take a look at the notebook (https://github.com/PacktPublishing/Applied-Machine-Learning-Explainability-Techniques/blob/main/Chapter02/Counterfactual_unstructured_data.ipynb) in the GitHub repository to learn how to implement this method for practical problems.

The following diagram, which has been taken from the paper *Counterfactual Visual Explanations, Goyal et al. 2019* (https://arxiv.org/pdf/1904.07451.pdf), shows how visual CFEs can be an effective approach for explaining image classifiers:

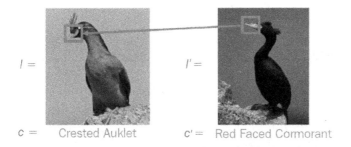

Figure 2.16 – A counterfactual example-based explanation for images

In practice, a CFE with unstructured data is difficult to achieve. It is still an area of active research, but I think this approach holds great potential to provide human-friendly explanations to even complex models. The following diagram shows the mapping of the various methods based on their explainability type:

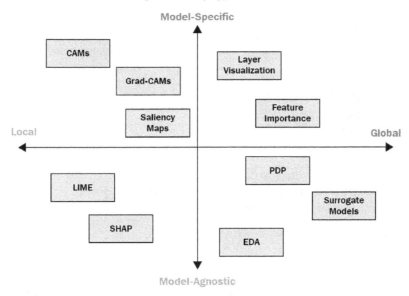

Figure 2.17 – Mapping various methods based on their explainability type

LIME and SHAP are important local and model-agnostic algorithms that are not covered in this chapter, but they will be covered in more detail later. The model explainability methods discussed in this chapter are widely used with a variety of datasets to provide different dimensions of explainability.

Summary

In this chapter, you learned about the various model explainability methods used to explain black-box models. Some of these are model-agnostic, while some are model specific. Some of these methods provide global interpretability, while some of them provide local interpretability. For most of these methods, visualizations through plots, graphs, and transformation maps are used to qualitatively inspect the data or the model outcomes; while for some of the methods, certain examples are used to provide explanations. Statistics and numerical metrics can also play an important role in providing quantitative explanations.

In the next chapter, we will discuss the very important concept of data-centric XAI and gain a conceptual understanding of how data-centric approaches can be leveraged in model explainability.

References

To gain additional information about the topics in this chapter, please refer to the following resources:

- *Friedman, Jerome H. "Greedy function approximation: A gradient boosting machine." Annals of statistics (2001)*: https://www.researchgate.net/publication/280687718_Greedy_Function_Approximation_A_Gradient_Boosting_Machine

- *Identifying outliers with the 1.5xIQR rule*: https://www.khanacademy.org/math/statistics-probability/summarizing-quantitative-data/box-whisker-plots/a/identifying-outliers-iqr-rule

- *Nelson rules*: https://www.leansixsigmadefinition.com/glossary/nelson-rules/

- *Accumulated Local Effects (ALE) – Feature Effects Global Interpretability*: https://www.analyticsvidhya.com/blog/2020/10/accumulated-local-effects-ale-feature-effects-global-interpretability/

- *Model-Agnostic Local Explanations using Individual Conditional Expectation (ICE) Plots*: https://towardsdatascience.com/how-to-explain-and-affect-individual-decisions-with-ice-curves-1-2-f39fd751546f

- *Figure 2.16: Counterfactual Visual Explanations, Goyal et al. 2019*: `https://arxiv.org/pdf/1904.07451.pdf`

- *Figure 2.13: Interpretability Beyond Feature Attribution: Quantitative Testing with Concept Activation Vectors (TCAV), Kim et al. [2018]*: `https://arxiv.org/abs/1711.11279`

- *Grad-CAM class activation visualization*: `https://keras.io/examples/vision/grad_cam/`

- *Generalized way of Interpreting CNNs using Guided Gradient Class Activation Maps!!*: `https://medium.com/@chinesh4/generalized-way-of-interpreting-cnns-a7d1b0178709`

- *Figure 2.12: Grad-CAM: Visual Explanations from Deep Networks via Gradient-based Localization, Ramprasaath et. al* - `https://arxiv.org/abs/1610.02391`

3
Data-Centric Approaches

In the *Defining explanation methods and approaches* section of *Chapter 1*, *Foundational Concepts of Explainability Techniques*, when we looked at the various dimensions of explainability, we discussed how data is one of the important dimensions. In fact, all **machine learning** (**ML**) algorithms depend on the underlying data being used.

In the previous chapter, we discussed various *model explainability methods*. Most of the methods discussed in *Chapter 2*, *Model Explainability Methods*, are model-centric. The concepts and ideas discussed were focused on making black-box models interpretable. But recently, the ML and AI communities have realized the core importance of data for any analysis or modeling purposes. So, more and more AI researchers are exploring new ideas and concepts around **data-centric AI**.

Since data plays a vital role in the model-building and prediction process, it is even more important for us to explain the functioning of any ML and AI algorithm with respect to the underlying data. From what I have observed from my experience in this field, the failure of any ML systems in production happens neither due to the poor choice of ML algorithm nor due to an inefficient model training or tuning process, but rather it occurs mostly due to inconsistencies in the underlying dataset. So, this chapter focuses on the concepts of **data-centric explainable AI** (**XAI**).

The goal of this chapter is to introduce you to the concepts of data-centric XAI. After reading this chapter, you will get to know about the various methods that can be performed to check the quality of the data, which might influence the model outcome. For production-level ML systems, the inference data can have issues related to its consistency and quality. So, monitoring these drifts is extremely important. Additionally, there can be external noise or perturbations affecting the data that can impact the model. So, these are some data-centric approaches that we will be discussing that are used for explaining ML models. In this chapter, the following main topics will be covered:

- Introduction to data-centric XAI
- Thorough data analysis and profiling processes
- Monitoring and anticipating drifts
- Checking adversarial robustness
- Measuring data forecastability

Now, let's dive in!

Technical requirements

Similar to *Chapter 2, Model Explainability Methods*, for this chapter, certain tutorial examples to implement some of the techniques to perform data-centric XAI in Python on certain interesting datasets have been provided. We will be using Python Jupyter notebooks to run the code and visualize the output throughout this book. The code and dataset resources for this chapter can be downloaded or cloned from the following GitHub repository: `https://github.com/PacktPublishing/Applied-Machine-Learning-Explainability-Techniques/tree/main/Chapter03`. Other important Python frameworks required to run the code will be mentioned in the notebooks with other relevant details to understand the code implementation of these concepts. Other important Python frameworks required to run the code will be mentioned in the notebooks along with other relevant details to understand the code implementation of these concepts. Please note that this chapter mainly focuses on providing a conceptual understanding of the topics covered. The Jupyter notebooks will help you gain the supplementary knowledge that is required to implement these concepts in practice. I recommend that you first read this chapter and then work on executing the Jupyter notebooks.

Introduction to data-centric XAI

Andrew Ng, one of the influential thought leaders in the field of AI and ML, has often highlighted the importance of using a systematic approach to build AI systems with high-quality data. He is one of the pioneers of the idea of **data-centric AI**, which focuses on developing systematic processes to develop models using clean, consistent, and reliable data, instead of focusing on the code and the algorithm. If the data is consistent, unambiguous, balanced, and available in sufficient quantity, this leads to faster model building, improved accuracy, and faster deployment for any production-level system.

Unfortunately, all AI and ML systems that exist in production today are not in alignment with the principles of data-centric AI. Consequently, there can be severe issues with the underlying data that seldom get detected but eventually lead to the failure of ML systems. That is why **data-centric XAI** is important to inspect and evaluate the quality of the data being used.

When we talk about explaining the functioning of any black-box model with respect to the data, we should inspect the *volume of the data*, the *consistency of the data* (particularly for supervised ML problems), and the *purity and integrity of the data*. Now, let's discuss these important aspects of data-centric XAI and understand why they are important.

Analyzing data volume

One of the classical problems of ML algorithms is the lack of generalization due to **overfitting**. Overfitting can be reduced by adding more data or by getting datasets of the appropriate volume to solve the underlying problem. So, the very first question we should ask about the data to explain the black-box model is "*Is the model trained on sufficient data?*" But for any industrial application, since data is very expensive, adding more data is not always feasible. So, the question should be "*How do we find out if the model was trained on sufficient data?*"

One way to inspect whether the model was trained on sufficient data is by training models with 70%, 80%, and 90% of the training dataset. If the model accuracy shows an increasing trend, with an increase in the volume of the data, that means the volume of the data can influence the model's performance. If the accuracy of the trained model, which has been trained on the entire training dataset, is low, then an increasing trend of model accuracy with increasing data volume indicates that the model is not trained on sufficient data. Therefore, more data is needed to make the model more accurate and generalized.

For production systems, if data is continuously flowing and there is no constraint on the availability of data, continuous training and monitoring of the models should be done on changing volumes of the data to understand and analyze its impact on the overall model performance.

Analyzing data consistency

Data consistency is another important factor to inspect when explaining ML models with respect to the data. One of the fundamental steps of analyzing data consistency is by understanding the distribution of the data. If the *data is not evenly distributed*, if there is a *class imbalance*, or if the *data is skewed toward a particular direction*, it is very likely that the model will be *biased* or less efficient for a particular segment of the data.

For production systems, it has been often observed that the inference data used in the production application might have some variance with the data used during training and validation. This phenomenon is known as **data drift**, and it refers to an unexpected change in the data structure or the statistical properties of the dataset, thus making the data corrupt and hampering the functioning of the ML system.

Data drift is very common for most real-time predictive models. This is simply because, in most situations, the data distribution changes over a period of time. This can happen due to a variety of reasons, for instance, if the system through which the data is being collected (for example, sensors) starts malfunctioning or needs to be recalibrated, then data drift can occur. Other external factors such as the surrounding temperature and surrounding noise can also introduce data drift. There can be a natural change in the relationship between the features that might cause data to drift. Consequently, if the training data is significantly different from the inference data, the model will make an error when predicting the outcome.

Now, sometimes, there can be a drift in the whole dataset or there can be a drift in one or more features. If there is a drift in a single feature, this is referred to as **feature drift**. There are multiple ways to detect feature drift such as the **Population Stability Index (PSI)** (`https://www.lexjansen.com/wuss/2017/47_Final_Paper_PDF.pdf`), **Kullback–Leibler Divergence (KL Divergence)** (`https://www.countbayesie.com/blog/2017/5/9/kullback-leibler-divergence-explained`), and **Wasserstein metric (Earth Movers Distance)** (`http://infolab.stanford.edu/pub/cstr/reports/cs/tr/99/1620/CS-TR-99-1620.ch4.pdf`). The application of some of these techniques has been demonstrated at `https://github.com/PacktPublishing/Applied-Machine-Learning-Explainability-Techniques/tree/main/Chapter03`. These statistical methods are popular ways to measure the distance between two data distributions. So, if the distance is significantly large, this is an indication of the presence of drift.

Apart from feature drift, **Label Drift** or **Concept Drift** can occur if the statistical properties of the target variable change over a period of time due to unknown reasons. However, overall data consistency is an important parameter for *root cause analysis inspection* when interpreting black-box models.

Analyzing data purity

The datasets used for practical industrial problems are never clean, even though most organizations spend a significant amount of time and investment on data engineering and data curation to drive a culture of *data-driven decision-making*. Yet, almost all practical datasets are messy and require a systematic approach to curation and preparation.

When we train a model, usually, we put our efforts into data preprocessing and preparation steps such as *finding duplicates or unique values*, *removing noise or unwanted values* from the data, *detecting outliers*, *handling missing values*, *handling features with mixed data types*, and even *transforming raw features or feature engineering* to get better ML models. On a high level, these methods are meant for removing impurities from the training data. But what if *a black-box ML model is trained on a dataset with less purity and, hence, performs poorly*?

That is why analyzing data purity is an important step in data-centric XAI. Aside from the data preprocessing and preparation methods mentioned earlier, there are other data integrity issues that exist as follows:

- **Label ambiguity**: For supervised ML problems, **label ambiguity** can be a very critical problem. If two or more instances of a dataset, which are very similar, have multiple labels, then this can lead to label ambiguity. Ambiguous labeling of the target variable can increase the difficulty of even domain experts classifying the samples correctly. Label ambiguity can be a very common problem, as, usually, labeled datasets are prepared by human subject-matter experts who are prone to *human error*.

- **Dominating features frequency change** (**DFCC**): Inspecting DFFC in the training and inference dataset is another parameter that can cause data integrity issues. In *Chapter 2, Model Explainability Methods*, when we discussed feature importance, we understood that not all features within the dataset are equally important, and some of the features have more influential power on the model's decision-making process. These are the dominating features in the dataset, and if the variance in the values of the dominating features in the training and the inference dataset is high, it is very likely that the model will make errors when predicting the outcome.

Other data purity issues, such as the introduction of a new label or new feature category, or out of bound values (or anomalies) for a particular feature in the inference set, can cause the failure of ML systems in production.

The following table shows certain important data purity checks that can be performed using the **Deepchecks Python framework**:

Status	Check	Condition
✓	Single Value in Column - Test Dataset	Does not contain only a single value for all columns
✓	Mixed Nulls - Test Dataset	Not more than 1 different null types for all columns
✓	Mixed Data Types - Test Dataset	Rare data types in all columns are either more than 10.00% or less than 1.00% of the data
✓	String Mismatch - Test Dataset	No string variants for all columns
✓	Data Duplicates - Test Dataset	Duplicate data is not greater than 0%
✓	String Length Out Of Bounds - Test Dataset	Ratio of outliers not greater than 0% string length outliers for all columns
✓	Special Characters - Test Dataset	Ratio of entirely special character samples not greater than 0.10% for all columns
✓	Label Ambiguity - Test Dataset	Ambiguous sample ratio is not greater than 0%

Figure 3.1 – Data purity checks using the Deepchecks framework

Data-centric XAI also includes other parameters that can be analyzed such as *adversarial robustness* (https://adversarial-ml-tutorial.org/introduction/), the *trust score comparison* (https://arxiv.org/abs/1805.11783), *covariate shift* (https://arxiv.org/abs/2111.08234), *data leakage between the training and validation datasets* (https://machinelearningmastery.com/data-leakage-machine-learning/), and *model performance sensitivity analysis* based on data alterations. All of these concepts apply to both tabular data and unstructured data such as images and text.

To explore practical ways of data purity analysis, you can refer to the Jupyter notebook at https://github.com/PacktPublishing/Applied-Machine-Learning-Explainability-Techniques/blob/main/Chapter03/Data_Centric_XAI_part_1.ipynb. We will discuss these topics later in the chapter.

Thorough data analysis and profiling process

In the previous section, you were introduced to the concept of data-centric XAI in which we discussed three important aspects of data-centric XAI: analyzing data volume, data consistency, and data purity. You might already be aware of some of the methods of data analysis and data profiling that we are going to learn in this section. But we are going to assume that we already have a trained ML model and, now, we are working toward explaining the model's decision-making process by adopting data-centric approaches.

The need for data analysis and profiling processes

In *Chapter 2, Model Explainability Methods*, when we discussed knowledge extraction using **exploratory data analysis (EDA)**, we discovered that this was a pre-hoc analysis process, in which we try to understand the data to form relevant hypotheses. As data scientists, these initial hypotheses are important as they allow us to take the necessary steps to build a better model. But let's suppose that we have a baseline trained ML model and the model is not performing as expected because it is not meeting the benchmark accuracy scores that were set.

Following the principles of model-centric approaches, most data scientists might want to spend more time in hyperparameter tuning, training for a greater number of epochs, feature engineering, or choosing a more complex algorithm. After a certain point, these methods will become limited and provide a very small boost to the model's accuracy. That is when data-centric approaches prove to be very efficient.

Data analysis as a precautionary step

By the principles of data-centric explainability approaches, at first, we try to perform a thorough analysis of the underlying dataset. We try to randomly reshuffle the data to create different training and validation sets and observe any overfitting or underfitting effects. If the model is overfitting or underfitting, clearly more data is required to generalize the model. If the available data is not sufficient in volume, there are ways to generate synthetic or artificial data. One such popular technique used for image classification is **data augmentation** (`https://research.aimultiple.com/data-augmentation/`). The **synthetic minority oversampling technique (SMOTE)** (`https://machinelearningmastery.com/smote-oversampling-for-imbalanced-classification/`) is also a powerful method that you can use to increase the size of the dataset. Some of these data-centric approaches are usually practiced during conventional ML workflows. However, we need to realize the importance of these steps for the explainability of black-box models.

Once we have done enough tests to understand whether the volume of the data is sufficient, we can try to inspect the consistency and purity of the data at a segmented level. If we are working on a classification problem, we can try to understand whether the model performance is consistent for all the classes. If not, we can isolate the particular class or classes for which the model performance is poor. Then, we check for data drifts, feature drifts, concept drifts, label ambiguity, data leakage (for example, when unseen test data trickles into the training data), and other data integrity checks for that particular class (or classes). Usually, if there is any abnormality with the data for a particular class (or classes), these checks are sufficient to isolate the problem. A thorough data analysis acts as a precautionary step to detect any loopholes in the modeling process.

Building robust data profiles

Another approach is to build statistical profiles of the data and then compare the profiles between the training data and the inference data. A statistical profile of a dataset is a collection of certain statistical measures of its feature values segmented by the target variable class (or, in the case of a regression problem, the bin of values). The selection of the statistical measures might change from use case to use case, but usually, I select statistical measures such as the mean, median, average variance, average standard deviation, coefficient of variation (standard deviation/mean), and z-scores ((value – mean)/standard deviation) for creating data profiles. In the case of time series data, measures such as the moving average and the moving median can also be very important.

Next, let's try to understand how this approach is useful. Suppose there is an arbitrary dataset that has three classes (namely class 0, 1, and 2) and only two features: feature 1 and feature 2. When we try to prepare the statistical profile, we would try to calculate certain statistical measures (such as the mean, median, and average variance in this example) for each feature and each class.

So, for class 0, a set of profile values consisting of the mean of feature 1, the median of feature 1, the average variance of feature 1, the mean of feature 2, the median of feature 2, and the average variance of feature 2 will be generated. Similarly, for class 1 and class 2, a set of profile values will be created for each class. The following table represents the statistical profile of the arbitrary dataset that we have considered for this example:

Class	Mean_feat1	Median_feat1	AvgVar_feat1	Mean_feat2	Median_feat2	AvgVar_feat2
0	34.5	37.0	-3.5	128.0	103.5	4.0
1	23.8	23.8	7.8	73.9	102.8	2.2
2	49.0	40	-2.3	101.5	101.5	-1.8

Figure 3.2 – A table showing a statistical profile segmented by each class for an arbitrary dataset

These statistical measures of the feature values can be used to compare the different classes. If a trained model predicts a particular class, we can compare the feature values with the statistical profile values for that particular class to get a fair idea about the influential features contributing to the decision-making process of the model. But more importantly, we can create separate statistical profiles for the validation set, test set, and inference data used in the production systems and compare them with the statistical profile of the training set. If the absolute percentage change between the values is significantly higher (say, > 20%), then this indicates the presence of data drift.

In our example, let's suppose that if the absolute percentage change in the average variance score for feature 1 for class 1 is about 25% between the training data and the inference data, then we have a feature drift for feature 1, and this might lead to poor model performance with the inference data for the production systems. Statistical profiles can also be created for unstructured data such as images and text, although the choice of statistical measures might be slightly complicated.

In general, this approach is very easy to implement and it helps us to validate whether the data used for training a model and the data used during testing or inference are consistent or not, which is an important step for data-centric model explainability. In the next section, we will discuss about the importance of monitoring and anticipating drifts for explaining ML systems.

Monitoring and anticipating drifts

In the previous section, we understood how a thorough data analysis and data profiling approach can help us to identify data issues related to volume, consistency, and purity. Usually, during the initial data exploration process, most data scientists try to inspect issues in the dataset in terms of volume and purity and perform necessary preprocessing and feature engineering steps to handle these issues.

But the detection of data consistency for real-time systems and production systems is a challenging problem for almost all ML systems. Additionally, issues relating to data consistency are often overlooked and are quite unpredictable as they can happen at any point in time in production systems. Some of the cases where data consistency issues can occur are listed as follows:

- They can occur due to natural reasons such as changes in external environmental conditions or due to the natural wear and tear of sensors or systems capturing the inference data.

- They can happen due to human-induced reasons such as any physical damage caused to the system collecting the data, any bug in the software program running the algorithm due to which the input data is being transformed incorrectly, or any noise introduced to the system while upgrading an older version of the system.

So, all of these situations can introduce data drifts and concept drifts, which eventually lead to the poor performance of ML models. And since drifts are very common in reality, issues related to drifts should be pre-anticipated and should be considered during the design process of any ML system.

Detecting drifts

After trained models are deployed for any production ML system, performance monitoring and feedback based on model performance is a necessary process. As we monitor the model performance, checking for any data or concept drifts is also critical in this step. At this point, you might be wondering two things:

- *What is the best way to identify the presence of a drift?*
- *What happens when we detect the presence of a drift?*

As discussed in the *Analyzing data consistency* section, there are two types of data drifts – *feature drifts* and *concept drifts*. Feature drifts happen when the statistical properties of the features or the independent variables change due to an unforeseen reason. In comparison, concept drift occurs when the target class variable, which the model is trying to predict, changes its initial relationship with the input features in a dynamic setting. In both cases, there is a statistical change in the underlying data. So, my recommendation for detecting drifts is to use the data profiling method discussed in the previous section.

A real-time monitoring dashboard is always helpful for any real-time application to monitor any drift. In the dashboard, try to have necessary visualizations for each class and each feature, comparing the statistical profile values with the actual live values flowing into the trained model.

Particularly for concept drifts, comparing the correlations of the features with the target outcome is extremely helpful. Since drifts can arise after a certain period of time or even during a specific point in time due to external reasons, it is always advisable to monitor the statistical properties of the inference data in a time window period (for instance, for 50 consecutive data points or 100 consecutive data points) rather than a continuous cumulative basis. For the purposes of feedback, necessary alerts and triggers can be set when abnormal data points are detected in the inference data, which might indicate the presence of data drift.

Selection of statistical measures

Sometimes, the selection of statistical measures can be difficult. So, we usually go for some popular distribution metrics to detect the presence of data drift using a quantitative approach. One such metric is called **trust score distribution** (https://arxiv.org/abs/1805.11783).

The following diagram shows the trust score distribution plot obtained using the *Deepchecks Python framework*:

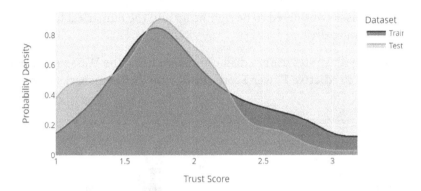

Figure 3.3 – An example of the trust score distribution between the training dataset
and the inference dataset

Trust score is a distribution metric used to measure the agreement between the ML classifier on the training set and an updated **k-Nearest Neighbor (kNN)** classifier on the inference dataset. The preceding diagram shows a trust score distribution plot between the training dataset and the inference dataset.

Ideally, the distributions should be almost the same for both the train and test datasets. However, if the trust score distribution for the inference set is skewed toward the extreme left, this indicates that the trained model has less confidence in the inference data, thereby alluding to the presence of drift. If the distribution of the trust score on the inference data is skewed toward the extreme right, there might be some problem with the model and there is a high probability of data leakage, as ideally, the trained model cannot be more confident in the test data in comparison to the training data.

To detect feature drifts on categorical features, the popular choice of metric is the **population stability index (PSI)** (https://www.lexjansen.com/wuss/2017/47_ Final_Paper_PDF.pdf). This is a statistical method used to measure the shift in a variable over a period of time. If the overall drift score is more than 0.2 or 20%, then the drift is considered to be significant, establishing the presence of feature drift.

To detect feature drifts in numeric features, the **Wasserstein metric** (`https://kowshikchilamkurthy.medium.com/wasserstein-distance-contraction-mapping-and-modern-rl-theory-93ef740ae867`) is the popular choice. This is a distance function for measuring the distance between two probability distributions. Similar to PSI, if the drift score using the Wasserstein metric is higher than 20%, this is considered to be significant and the numerical feature is considered to have feature drift.

The following diagram illustrates feature drift estimation using the Wasserstein (Earth Mover's) distance and **Predictive Power Score** (**PPS**) with the Deepchecks framework:

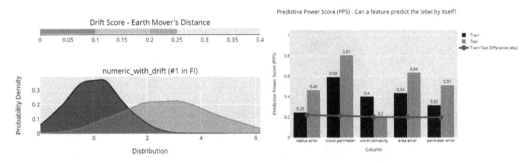

Figure 3.4 – Feature drift estimation using Wasserstein (Earth Mover's) distance and PPS of features

Similar concept drifts can also be detected using these metrics. For regression problems, the Wasserstein metric is effective, while for classification problems PSI is more effective. You can see the application of these methods on a practical dataset at `https://github.com/PacktPublishing/Applied-Machine-Learning-Explainability-Techniques/tree/main/Chapter03`. Additionally, there are other statistical methods that are extremely useful for detecting data drifts such as **Kullback-Leibler Divergence** (**KL Divergence**), the **Bhattacharyya distance**, **Jensen-Shannon Divergence** (**JS Divergence**), and more.

In this chapter, our focus is not on learning these metrics, but I strongly recommend you to take a look at the *Reference* section to find out more about these metrics and their application for finding data drifts. These methods are also applicable to images. Instead of structured feature values, the distributions of the pixel intensity value of the image datasets are used to detect drifts.

Now that we are aware of certain effective ways in which to detect drifts, *what do we do when we have identified the presence of drifts?* The first step is to alert our stakeholders if the ML system is already in production. Incorrect predictions due to data drift can impact many end users, which might, ultimately, lead to the loss of trust of the end users. The next step is to check whether the drift is *temporary*, *seasonal*, or *permanent* in nature. Analysis of the nature of the drift can be challenging, but if the changes that are causing the drift can be identified and reverted, then that is the best solution.

If the drift is temporary, the first step is to identify the temporary change that caused the drift and then revert the changes. For seasonal drifts, seasonal changes to the data should be accounted for during the training process or as an additional preprocessing step to normalize any seasonal effects on the data. This is so that the model is aware of the seasonal pattern in the data. However, if the drift is permanent, then the only option is to retrain the model on the new data and deploy the newly trained model for the production system.

In the context of XAI, the detection of drifts can justify the failure of any ML model or algorithm and helps to improve the model by identifying the root cause of the failure. In the next section, we will discuss another data-centric quality inspection step that can be performed to ensure the robustness of ML models.

Checking adversarial robustness

In the previous section, we discussed the importance of anticipating and monitoring drifts for any production-level ML system. Usually, this type of monitoring is done after the model has been deployed in production. But even before the model is deployed in production, it is extremely critical to check for the **adversarial robustness** of the model.

Most ML models are prone to adversarial attacks or an injection of noise to the input data, causing the model to fail by making incorrect predictions. The degree of adversarial attacks increases with the model's complexity, as complex models are very sensitive to noisy data samples. So, checking for adversarial robustness is about evaluating how sensitive the trained model is toward adversarial attacks.

In this section, first, we will try to understand the impact of adversarial attacks on the model and why this is important in the context of XAI. Then, we will discuss certain techniques that we can use to increase the adversarial robustness of ML models. Finally, we will discuss the methods that are used to evaluate the adversarial robustness of models, which can be performed as an exercise before deploying ML models into production, and how this forms a vital part of explainable ML systems.

Impact of adversarial attacks

Over the past few years, adversarial attacks have been a cause of great concern for the AI community. These attacks can inject noise to modify the input data in such a way that a human observer can easily identify the correct outcome but an ML model can be easily fooled and start predicting completely incorrect outcomes. The extent of the attack depends on the attacker's access to the model.

Usually, in production systems, the trained model (especially the model parameters) is fixed and cannot be modified. But the inference data flowing into the model can be polluted with abrupt noise signals, thus making the model misclassify. Human experts are extremely efficient in filtering out the injected noise, but ML models fail to isolate the noise from the actual data if the model has not been exposed to such noisy samples during the training phase. Sometimes, these attacks can be *targeted* attacks, too.

For example, if a face recognition system allows access to only a specific person, adversarial attacks can modify the image of any person to a specific person by introducing some noise. In this case, an adversarial algorithm would have to be trained using the target sample to construct the noise signal. There are other forms of adversarial attacks as well, which can modify the model during the training phase itself. However, since we are discussing this in the context of XAI, we will concentrate on the impact of adversarial effects on trained ML models.

There are different types of adversarial attacks that can impact trained ML models:

- **Fast Gradient Sign Method (FGSM)**: FGSM is one such method that uses gradients of deep learning models to learn adversarial samples. For image classifiers, this can be a common problem, as FGSM creates perturbations on the pixel values of an image by adding or subtracting pixel intensity values depending on the *direction of the gradient descent* of the model. This can fool the model to misclassify and severely affect the performance of the model, but it does not create any problem for a human observer. Even if the modification appears to be negligible, the method adds an evenly distributed noise that is enough to cause the misclassifications.

- **The Carlini & Wagner (C&W) attack**: Another common adversarial attack is the C&W attack. This method uses the three norm-based distance metrics (L_0, L_2, and L_∞) to find adversarial examples, such that the distance between the adversarial example and the original sample is minimal. Detecting C&W attacks is more difficult than FGSM attacks.

- **Targeted adversarial patch attacks**: Sometimes, injecting noise (that is, the addition of noisy random pixels) into the entire image is not necessary. The addition of a noisy image segment to only a small portion of the image can be equally harmful to the model. Targeted adversarial patch attacks can generate a small adversarial patch that is then superimposed with the original sample, thus occluding the key features of the data and making the model classify incorrectly. There are other forms of adversarial attacks too, and many more new methods can be discovered in the future. However, the impact will still be the same.

The following diagram shows how different adversarial attacks can introduce noise in an image, thereby making it difficult for the model to give correct predictions. Despite the addition of noise, we, as human beings, can still predict the correct outcome, but the trained model is completely fooled by adversarial attacks:

Figure 3.5 – Adversarial attacks on the inference data leading to incorrect model predictions

Adversarial attacks can force an ML model to produce incorrect outcomes that can severely affect end users. In the next section, let's try to explore ways to increase the adversarial robustness of models.

Methods to increase adversarial robustness

In production systems, adversarial attacks can mostly inject noise into the inference data. So, to reduce the impact of adversarial attacks, we would either need to teach the model to filter out the noise or expose the presence of noisy samples during the training process or train the models to detect adversarial samples:

- The easiest option is to filter out the noise as a defense mechanism to increase the adversarial robustness of ML models. Any adversarial noise results in an abrupt change in the input samples. In order to filter out any abrupt change from any signal, we usually try to apply a smoothing filter such as **Spatial smoothing**. Spatial smoothing is equivalent to the **blurring operation** in images and is used to reduce the impact of the adversarial attack. From experience, I have observed that an *adaptive median spatial smoothing* (https://homepages.inf.ed.ac.uk/ rbf/HIPR2/median.htm), which works at a local level through a windowing approach is more effective than smoothing at a global level. The statistical measure of the median is always more effective in filtering out noise or outliers from the data.

- Another approach to increase adversarial robustness is by introducing adversarial examples during the training process. By using the technique of **data augmentation**, we can generate adversarial samples from the original data and include the augmented data during the training process. If training the model from scratch using augmented adversarial samples is not feasible, then the trained ML model can actually be fine-tuned on the adversarial samples using **transfer learning**. Here, the trained model weights can be taken to be fine-tuned on the newer samples.

- The process of training a model with adversarial samples is often referred to as **adversarial training**. We can even train a separate model using adversarial training, just to detect adversarial samples from original samples, and use it along with the main model to trigger alerts if adversarial samples are generated. The idea of exposing the model to possible adversarial samples is similar to the idea of *stress testing* in cyber security (https://ieeexplore.ieee.org/ document/6459909).

Figure 3.6 illustrates how spatial smoothing can be used as a defense mechanism to minimize the impact of adversarial attacks:

Figure 3.6 – Using spatial smoothing as a defense mechanism to minimize
the impact of adversarial attacks

Using the methods that we have discussed so far, we can increase the adversarial robustness of trained ML models to a great extent. In the next section, we will try to explore ways to evaluate the adversarial robustness of ML models.

Evaluating adversarial robustness

Now that we have learned certain approaches in which to defend against adversarial attacks, the immediate question that might come to mind is *how can we measure the adversarial robustness of models?*

Unfortunately, I have never come across any dedicated metric to quantitatively measure the adversarial robustness of ML models, but it is an important research topic for the AI community. The most common approaches by which data scientists evaluate the adversarial robustness of ML models are **stress testing** and **segmented stress testing**:

- In **stress testing**, adversarial examples are generated by FGSM or C&W methods. Following this, the model's accuracy is measured on the adversarial examples and compared to the model accuracy obtained with the original data. The strength of the adversarial attack can also be increased or decreased to observe the variation of the model performance with the attack strength. Sometimes, a particular class or feature can become more vulnerable to adversarial attacks than the entire dataset. In those scenarios, segmented stress testing is beneficial.

- In **segmented stress testing**, instead of measuring the adversarial robustness of the entire model on the entire dataset, segments of the dataset (either for specific classes or for specific features) are considered to compare the model robustness with the adversarial attack strengths. Adversarial examples can be generated with random noise or with Gaussian noise. For certain datasets, quantitative metrics such as the **Peak Signal-to-Noise ratio** (**PSNR**) (`https://www.ni.com/nl-be/innovations/white-papers/11/peak-signal-to-noise-ratio-as-an-image-quality-metric.html`) and **Erreur Relative Globale Adimensionnelle de Synthese** (**ERGAS**) (`https://www.researchgate.net/figure/Erreur-Relative-Globale-Adimensionnelle-de-Synthese-ERGAS-values-of-fused-images_tbl1_248978216`) are used to measure the data or signal quality. Otherwise, the adversarial robustness of ML models can be quantitatively inspected by the model's prediction of adversarial samples.

More than the method of evaluating adversarial robustness, inspecting adversarial robustness and monitoring the detection of adversarial attacks is an essential part of explainable ML systems. Next, let's discuss the importance of measuring data forecastability as a method to provide data-centric model explainability.

Measuring data forecastability

So far, we have learned about the importance of analyzing data by inspecting its consistency and purity, looking for monitoring drifts, and checking for any adversarial attacks to explain the working of ML models. But some datasets are extremely complex and, hence, training accurate models even with complex algorithms is not feasible. If the trained model is not accurate, it is prone to make incorrect predictions. Now the question is *how do we gain the trust of our end users if we know that the trained model is not extremely accurate in making the correct predictions?*

I would say that the best way to gain trust is by being transparent and clearly communicating what is feasible. So, measuring **data forecastability** and communicating the model's efficiency to end users helps to set the right expectation.

Data forecastability is an estimation of the model's performance using the underlying data. For example, let's suppose we have a model to predict the stock price of a particular company. The stock price data that is being modeled by the ML algorithm can predict the stock price with a maximum of 60% accuracy. Beyond that point, it is not practically possible to generate a more accurate outcome using the given dataset.

But let's say that if other external factors are considered to supplement the current data, the model's accuracy can be boosted. This proves that it is not the ML algorithm that is limiting the performance of the system, but rather the dataset that is used for modeling does not have sufficient information to get a better model performance. Hence, it is a limitation of the dataset that can be estimated by measure of data forecastability.

The following diagram shows a number of model evaluation visualizations that can be used to analyze data forecastability using the Deepchecks framework:

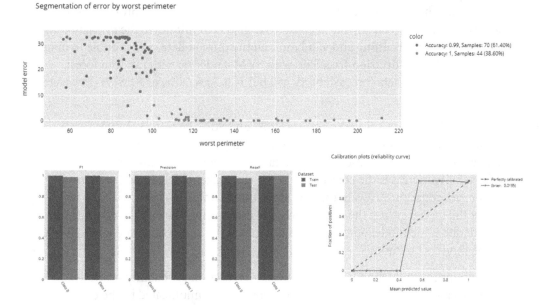

Figure 3.7 – Data forecastability using the model evaluation report and the Deepchecks framework

Next, let's discuss how to estimate data forecastability.

Estimating data forecastability

Data forecastability is estimated using the model evaluation metrics. Data forecastability can also be measured by performing model error analysis. The choice of the metrics depends on the type of dataset and the type of problem being solved. For example, take a look at the following list:

- For time series data, data forecastability is obtained by metrics such as the **mean absolute percentage error (MAPE)**, the **symmetric mean absolute percentage error (SMAPE)**, the **coefficient of variation (CoV)**, and more.

- For classification problems, I usually go for **ROC-AUC Scores**, the **confusion matrix**, **precision**, **recall**, and **F1 scores** along with **accuracy**.

- For regression problems, we can look at the **mean square error (MSE)**, the **R2 score**, the **root mean square error (RMSE)**, the **sum of squared errors (SSE)**, and more.

You might have already used most of these metrics to evaluate trained ML models. Data forecastability is not just about evaluating trained models according to your choice of metric, but is the measure of predictability of the model using the given dataset.

Let's suppose you are applying three different ML algorithms such as decision trees, support vector machine (SVM), and random forests for a classification problem, and your choice of metric is recall. This is because your goal is to minimize the impact of false positives. After rigorous training and validation on the unseen data, you are able to obtain recall scores of 70% with decision tree, 85% with SVM, and 90% with random forest. What do you think your data forecastability will be? Is it 70%, 90%, or 81.67% (the average of the three scores)?

I would say that the correct answer is between 70% and 90%. It is always better to consider forecastability as a ballpark estimate, as providing a range of values rather than a single value gives an idea of the best-case and worst-case scenarios. Communicating about the data forecastability increases the confidence of the end stakeholders in ML systems. If the end users are consciously aware that the algorithm is only 70% accurate, they will not blindly trust the model even if the system predicts incorrectly. The end users would be more considerate if the model outcome does not match the actual outcome when they are aware of the model's limitations.

Most ML systems in production have started using prediction probability or model confidence as a measure of data forecastability, which is communicated to the end users. For example, nowadays, most weather forecasting applications show that there is a certain percentage of chance (or probability) for rainfall or snowfall. Therefore, data forecastability increases the explainability of AI algorithms by setting up the right expectation for the accuracy of the predicted outcome. It is not just the measure of the model performance, but rather a measure of the predictability of a model which is trained on a specific dataset.

This brings us to the end of the chapter. Let's summarize what we have discussed in the following section.

Summary

Now, let's try to summarize what you have learned in this chapter. In this chapter, we focused on data-centric approaches for XAI. We learned the importance of explaining black-box models with respect to the underlying data, as data is the central part of any ML model. The concept of data-centric XAI might be new to many of you, but it is an important area of research for the entire AI community. Data-centric XAI can provide explainability to the black-box model in terms of data volume, data consistency, and data purity.

Data-centric explainability methods are still active research topics, and there is no single Python framework that exists that covers all of the various aspects of data-centric XAI. Please explore the supplementary Jupyter notebook tutorials provided at `https://github.com/PacktPublishing/Applied-Machine-Learning-Explainability-Techniques/tree/main/Chapter03` to gain more practical knowledge on this topic.

We learned about the idea of thorough data inspection and data profiling to estimate the consistency of training data and inference data. Monitoring data drifts for production ML systems is also an essential part of the data-centric XAI process. Apart from data drifts, estimating the adversarial robustness of ML models and the detection of adversarial attacks form an important part of the process.

Finally, we learned about the importance of data forecastability to set the right expectation to end stakeholders about what the model can achieve and how this is a necessary practice that can increase the trust of our end users.

You have been introduced to many statistical concepts in this chapter. Covering everything about each statistical method is beyond the scope of this chapter. However, I strongly recommend that you go through the reference links shared to understand these topics in greater depth.

This brings us to the end of part 1 of this book, in which you have been exposed to the conceptual understanding of certain key topics of XAI. From the next chapter onward, we will start exploring popular Python frameworks for applying the concepts of XAI to practical real-world problems. In the next chapter, we will cover an important XAI framework called LIME and examine how it can be used in practice.

References

To gain additional information about the topics in this chapter, please refer to the following resources:

- *Andrew Ng Launches A Campaign For Data-Centric AI*: `https://www.forbes.com/sites/gilpress/2021/06/16/andrew-ng-launches-a-campaign-for-data-centric-ai/?sh=5333db3a74f5`

- *Landing.AI: Data-Centric AI*: `https://landing.ai/data-centric-ai/`

- *Jiang et al. "To Trust Or Not To Trust A Classifier" (2018)*: `https://arxiv.org/abs/1805.11783`

- *Deepchecks*: `https://docs.deepchecks.com/en/stable/`

- *Statistical distance*: `https://en.wikipedia.org/wiki/Statistical_distance`

- *Lin et al. "Examining Distributional Shifts by Using Population Stability Index (PSI) for Model Validation and Diagnosis"*: `https://www.lexjansen.com/wuss/2017/47_Final_Paper_PDF.pdf`

- *Wasserstein metric*: `https://en.wikipedia.org/wiki/Wasserstein_metric`

- *Bhattacharyya Distance based Concept Drift Detection Method For evolving data stream*: `https://www.researchgate.net/publication/352044688_Bhattacharyya_Distance_based_Concept_Drift_Detection_Method_For_evolving_data_stream`

Section 2 – Practical Problem Solving

This section will give you the information and experience needed to implement the various approaches to model explainability using Python. You will learn about the different Python frameworks for implementing the concepts of **Explainable AI (XAI)** covered in the previous section, such as LIME, SHAP, TCAV, ALIBI, DALEX, Explainerdashboard, InterpretML, ELI5, and DiCE. You will also get the necessary practical exposure to apply explainability methods for practical use cases.

This section comprises the following chapters:

- *Chapter 4, LIME for Model Interpretability*
- *Chapter 5, Practical Exposure to Using LIME in ML*
- *Chapter 6, Model Interpretability Using SHAP*
- *Chapter 7, Practical Exposure to Using SHAP in ML*
- *Chapter 8, Human-Friendly Explanations with TCAV*
- *Chapter 9, Other Popular XAI Frameworks*

4
LIME for Model Interpretability

In the previous chapters, we discussed the various technical concepts of **Explainable AI** (**XAI**) that are needed to build trustworthy AI systems. Additionally, we looked at certain practical examples and demonstrations using various Python frameworks to implement the concepts of practical problem solving, which are given in the GitHub code repository of this chapter. XAI has been an important research topic for quite some time, but it is only very recently that all organizations have started to adopt XAI as a part of the solution life cycle for solving business problems using AI. One such popular approach is **Local Interpretable Model-Agnostic Explanations** (**LIME**), which has been widely adopted to provide model-agnostic local explainability. The LIME Python library is a robust framework that provides human-friendly explanations to tabular, text, and image data and helps in interpreting black-box supervised machine learning algorithms.

In this chapter, you will be introduced to the LIME framework, which has made a significant impact in the field of XAI. We will discuss the workings of the LIME algorithm for global and local model explainability. Also, I will demonstrate an example in which the LIME Python framework can be used in practice. I will cover the limitations of this framework that you should be aware of.

So, in this chapter, we will discuss the following main topics:

- An intuitive understanding of LIME

- What makes LIME a good model explainer?

- Submodular pick (SP-LIME)

- A practical example of using LIME for classification problems

- Potential pitfalls

Without further ado, let's get started.

Technical requirements

This chapter is slightly more technical than the previous chapters covered in this book. The code and dataset resources can be downloaded or cloned from the GitHub repository for this chapter, which is located at `https://github.com/PacktPublishing/ Applied-Machine-Learning-Explainability-Techniques/tree/main/ Chapter04`. Similar to the previous chapters, we will be using Python and Jupyter notebooks to run the code and generate the necessary outputs. Other important Python frameworks that are necessary to run the code will be mentioned in the notebooks with further relevant details to understand the code implementation of these concepts.

Intuitive understanding of LIME

LIME is a novel, model-agnostic, local explanation technique used for interpreting black-box models by learning a local model around the predictions. LIME provides an intuitive global understanding of the model, which is helpful for non-expert users, too. The technique was first proposed in the research paper *"Why Should I Trust You?" Explaining the Predictions of Any Classifier* by *Ribeiro et al.* (`https://arxiv.org/ abs/1602.04938`). The Python library can be installed from the GitHub repository at `https://github.com/marcotcr/lime`. The algorithm does a pretty good job of interpreting any classifier or regressor in faithful ways by using approximated local interpretable models. It provides a global perspective to establish trust for any black-box model; therefore, it allows you to identify interpretable models over human-interpretable representation, which is locally faithful to the algorithm. So, it mainly functions by *learning interpretable data representations, maintaining a balance in a fidelity-interpretability trade-off,* and *searching for local explorations.* Let's look at each one of them in detail.

Learning interpretable data representations

LIME does a pretty good job in differentiating between impactful features and choosing interpretable data representations that are understandable to any non-expert user regardless of the actual complex features used by the algorithm. For example, when explaining models trained on unstructured data such as images, the actual algorithm might use complex numerical feature vectors for its decision-making process, but these numerical feature values are incomprehensible to any non-technical end user. In comparison, if the explainability is provided in terms of the presence or absence of a region of interest or superpixel (that is, a continuous patch of pixels) within the image, that is a human-interpretable way of providing explainability.

Similarly, for text data, instead of using word-embedding vector values to interpret models, a better way to provide a human-interpretable explanation is by using examples of the presence or absence of certain words used to describe the target outcome of the model. So, mathematically speaking, the original representation of a data instance being explained is denoted by $x \in R^d$, where d is the entire dimension of data. A binary vector of interpretable data representations is mathematically denoted by $x' \epsilon \{0,1\}^{d'}$. Intuitively speaking, the algorithm tries to denote the presence or absence of human-interpretable data representations to explain any black-box model.

Figure 4.1 shows how LIME tries to divide the input image data into human-interpretable components that are later used to explain black-box models in a manner that is understandable to any non-technical user:

Figure 4.1 – How LIME transforms an image into human-interpretable components

Next, let's discuss how to maintain the fidelity-interpretability trade-off.

Maintaining a balance in the fidelity-interpretability trade-off

LIME makes use of inherently interpretable models such as decision trees, linear models, and rule-based heuristic models to provide explanations to non-expert users with visual or textual artifacts. Mathematically speaking, this explanation is a model that can be denoted by $g \in G$, where G is the entire set of potentially interpretable models and the domain of g is represented with another binary vector, $\{0,1\}^{d'}$, which represents the presence or absence of interpretable components. Additionally, the algorithm tries to measure the *complexity* of an explanation along with its *interpretability*. For example, even in interpretable models such as decision trees, the depth of the tree is a measure of its complexity.

Mathematically speaking, the complexity of an interpretable model is denoted by $\Omega(g)$. LIME tries to maintain **local fidelity** while providing explanations. This means that the algorithm tries to replicate the behavior of the model in proximity to the individual data instance being predicted. So, mathematically, the inventors of this algorithm used a function, $\pi_x(z)$, to measure the proximity between any data instances, z, thus defining the locality around the original representation, x. Now, if the probability function, $f(x)$, defines the probability that x belongs to a certain class, then to approximate f, the LIME algorithm tries to measure how unfaithful g is with a proximity function, π_x. This entire operation is denoted by the $L(f, g, \pi_x)$ function. Therefore, the algorithm tries to minimize the locality-aware loss function, $L(f, g, \pi_x)$, while maintaining $\Omega(g)$ to be a low value. This is so that it is easily explainable to any non-expert user. The measure of an interpretability local fidelity trade-off is approximated by the following mathematical function:

$$\xi(x) = \operatorname*{argmin}_{g \in G} \, L(f, g, \pi_x) + \Omega(g)$$

Hence, this trade-off measure depends on the interpretable models, G, the fidelity function, L, and the complexity measure, Ω.

Searching for local explorations

The LIME algorithm is *model-agnostic*. This means when we try to minimize the *locality-aware loss function*, $L(f, g, \pi_x)$, without any assumption about f. Also, LIME maintains local fidelity by taking samples that are weighted by π_x while approximating $L(f, g, \pi_x)$. Nonzero samples of X' are drawn uniformly at random to sample instances around X'. Let's suppose there is a perturbed sample containing fractions of nonzero elements of X', which is denoted by $z' \in \{0,1\}^{d'}$. The algorithm tries to recover samples from the original representation, $z \in R^d$, to approximate $f(x)$. Then, $f(x)$ is used as a label for the explanation model, $\xi(x)$.

Figure 4.2 represents an example presented in the original paper of the LIME framework at `https://arxiv.org/pdf/1602.04938.pdf`, which intuitively explains the working of the algorithm using a visual representation:

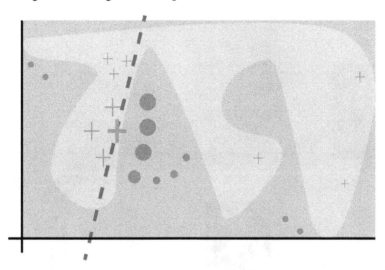

Figure 4.2 – Explaining the working of the LIME algorithm intuitively

In *Figure 4.2*, the curve separating the light blue and pink backgrounds is considered a complex f decision function of a black-box model. Since the decision function is not linear, approximating it using linear models is not efficient. The crosses and the dots represent training data belonging to two different classes. The bold cross represents the inference data instance being explained. The algorithm functions by sampling instances to get predictions using f. Then, the algorithm assigns weight by the proximity to the data instance being explained. In the preceding diagram, based on the proximity of the data instance, the sizes of the red crosses and blue dots are varied. So, the instances that are sampled are both in closer proximity to x, having a higher weight from π_x, and far away from it, thus having a lower weight of π_x. The original black-box model might be too complex to provide a global explanation, but the LIME framework can provide explanations that are appropriate for the local data instance, π_x. The learned explanation is illustrated by the dashed line, which is locally faithful with a global perspective.

Figure 4.3 illustrates a far more intuitive understanding of the LIME algorithm. From the original image, the algorithm generates a set of perturbed data instances:

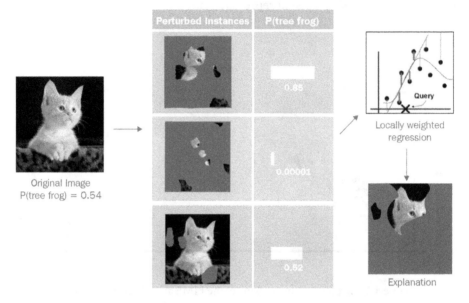

Figure 4.3 – Predictions being explained using LIME

The perturbed instances, as shown in *Figure 4.3*, are created by switching some of the interpretable components off. In the case of images, as shown in the preceding diagram, it is done by turning certain components gray. Then, the black-box model is applied to each of the perturbed instances that are generated, and the probability of the instance being predicted as the final outcome of the model is calculated. Then, an interpretable model (such as a simple locally weighted linear model) is learned on the dataset, and finally, the superpixels having the maximum positive weights are considered for the final explanation.

In the next section, let's discuss why LIME is a good model explainer.

What makes LIME a good model explainer?

LIME enables non-expert users to understand the working of untrustworthy black-box models. The following properties of LIME make it a good model explainer:

- **Human interpretable**: As discussed in the previous section, LIME provides explanations that are easy to understand, as it provides a qualitative way to compare the components of the input data with the model outcome.

- **Model-agnostic**: In the previous chapters, although you have learned about various model-specific explanation methods, it is always an advantage if the explanation method can be used to provide explainability for any black-box model. LIME does not make any assumptions about the model while providing the explanations and can work with any model.

- **Local fidelity**: LIME tries to replicate the behavior of the entire model by exploring the proximity of the data instance being predicted. So, it provides local explainability to the data instance being used for prediction. This is important for any non-technical user to understand the exact reason for the model's decision-making process.

- **Global intuition**: Although the algorithm provides local explainability, it does try to explain a representative set to the end users, thereby providing a global perspective to the functioning of the model. SP-LIME provides a global understanding of the model by explaining a collection of data instances. This will be covered in more detail in the next section.

Now that we understand the key advantages of the LIME framework, in the next section, let's discuss the submodular pick algorithm of LIME, which is used for extracting global explainability.

SP-LIME

In order to make explanation methods more trustworthy, providing an explanation to a single data instance (that is, a local explanation) is not always sufficient, and the end user might want a global understanding of the model to have higher reliability on the robustness of the model. So, the SP-LIME algorithm tries to run the explanations on multiple diverse, yet carefully selected, sets of instances and returns non-redundant explanations.

Now, let me provide an intuitive understanding of the SP-LIME algorithm. The algorithm considers that the time required to go through all the individual local explanations is limited and is a constraint. So, the number of explanations that the end users are willing to examine to explain a model is the budget of the algorithm denoted by B. Let's suppose that X denotes the set of instances; the task of selecting B instances for the end user to analyze for model explainability is defined as the **pick step**. The pick step is independent of the existence of the explanation and it needs to provide *non-redundant explanations* by picking up a diverse representative set of instances to explain how the model is behaving considering a global perspective. Therefore, the algorithm tries to avoid picking up instances with similar explanations.

Mathematically, this idea is represented using the *Explanation Matrix* (*W*), in which *W* = *n* * *d'*, such that *n* is the number of samples and *d'* is the human interpretable features. The algorithm also uses a *Global importance component matrix* (*I*), in which for each component of *j*, *I(j)* represent the global importance in the explanation space. Intuitively speaking, *I* is formulated in a way to assign higher scores to features, which explains many instances of the data. The set of important features that are considered for the explanations is denoted by *V*. So, combining all these parameters, the algorithm tries to learn a *non-redundant coverage intuition function, c(V,W,I)*. The non-redundant coverage intuition tries to compute the collective importance of all features that appear in at least one instance in set *V*. However, the *pick problem* is about *maximizing the weighted coverage function*. This is denoted by the following equation:

$$Pick(W,I) = \underset{V,|V|\leq B}{\text{argmax}}\, c(V,W,I)$$

The details about the algorithm that we just covered in this section might be slightly overwhelming to understand for certain readers. However, intuitively, the algorithm tries to cover the following steps:

1.   The explanation model is run on all instances (*x*).

2.   The global importance of all individual components is computed.

3.   Then, the algorithm tries to maximize the non-redundant coverage intuition function (*c*) by iteratively adding instances with the highest maximum coverage gain.

4.   Finally, the algorithm tries to obtain the representative non-redundant explanation set (*V*) and return it.

In the next section, we will cover how the LIME Python framework can be used for classification problems using code examples.

A practical example of using LIME for classification problems

So far, we have covered most of the in-depth conceptual understanding that is needed regarding the LIME algorithm. In this section, we will try to explore the LIME Python framework for explaining classification problems. The framework is available as an open source project on GitHub at `https://github.com/marcotcr/lime`. Installing LIME in Python can be done easily using the `pip` installer inside the Jupyter notebook:

```
!pip install lime
```

The complete notebook version of the tutorial is accessible from the GitHub repository at `https://github.com/PacktPublishing/Applied-Machine-Learning-Explainability-Techniques/blob/main/Chapter04/Intro_to_LIME.ipynb`. However, for now, I will try to walk you through the entire code so that you understand the code in detail. Once the LIME framework has been installed, quickly verify whether the installation was successful or not by importing the library:

```
import lime
```

If the import was successful, you can easily proceed with the next steps; otherwise, you need to check what went wrong while installing the framework. But usually, you should not face any errors or any dependency conflicts as installing the library is quite straightforward. For this tutorial, we will use the *Titanic dataset* (`https://www.openml.org/search?type=data&sort=runs&id=40945&status=active`). This is one of the classic machine learning datasets used for predicting the survival of passengers on the Titanic. So, this is a binary classification problem that can be solved using machine learning. Although this is a classic dataset that is not very complex, it contains all types of features such as *Categorical*, *Ordinal*, *Continuous*, and even certain *identifiers* that are not relevant for the classification, thereby making this an interesting dataset to work with. To make it easier for you to execute notebooks, I have downloaded and provided the dataset after some slight modifications in the code repository at `https://github.com/PacktPublishing/Applied-Machine-Learning-Explainability-Techniques/tree/main/Chapter04/dataset`.

Titanic dataset

The original Titanic dataset, describing the survival status of individual passengers on the Titanic. The titanic data does not contain information from the crew, but it does contain actual ages of half of the passengers. The principal source for data about Titanic passengers is the Encyclopedia Titanica. The datasets used here were begun by a variety of researchers. One of the original sources is Eaton & Haas (1994) Titanic: Triumph and Tragedy, Patrick Stephens Ltd, which includes a passenger list created by many researchers and edited by Michael A. Findlay.

Thomas Cason of UVa has greatly updated and improved the titanic data frame using the Encyclopedia Titanica and created the dataset here. Some duplicate passengers have been dropped, many errors corrected, many missing ages filled in, and new variables created.

After installing and importing all the required modules, first, we will start by loading the dataset from the directory as a pandas DataFrame:

```
data = pd.read _ csv('dataset/titanic.csv')
```

When you try to visualize the DataFrame using the `head` method from pandas, you will get a glimpse of the dataset, as shown in *Figure 4.4*. Often, this step helps you to get a quick idea about how to understand your data:

```
data.head()
```

The following diagram shows a glimpse of the pandas DataFrame used for this example:

PassengerId	Survived	Pclass	Name	Sex	Age	SibSp	Parch	Ticket	Fare	Cabin	Embarked
1	0	3	Braund, Mr. Owen Harris	male	22.0	1	0	A/5 21171	7.2500	NaN	S
2	1	1	Cumings, Mrs. John Bradley (Florence Briggs Th...	female	38.0	1	0	PC 17599	71.2833	C85	C
3	1	3	Heikkinen, Miss. Laina	female	26.0	0	0	STON/O2. 3101282	7.9250	NaN	S
4	1	1	Futrelle, Mrs. Jacques Heath (Lily May Peel)	female	35.0	1	0	113803	53.1000	C123	S
5	0	3	Allen, Mr. William Henry	male	35.0	0	0	373450	8.0500	NaN	S
6	0	3	Moran, Mr. James	male	NaN	0	0	330877	8.4583	NaN	Q
7	0	1	McCarthy, Mr. Timothy J	male	54.0	0	0	17463	51.8625	E46	S
8	0	3	Palsson, Master. Gosta Leonard	male	2.0	3	1	349909	21.0750	NaN	S
9	1	3	Johnson, Mrs. Oscar W (Elisabeth Vilhelmina Berg)	female	27.0	0	2	347742	11.1333	NaN	S
10	1	2	Nasser, Mrs. Nicholas (Adele Achem)	female	14.0	1	0	237736	30.0708	NaN	C
11	1	3	Sandstrom, Miss. Marguerite Rut	female	4.0	1	1	PP 9549	16.7000	G6	S
12	1	1	Bonnell, Miss. Elizabeth	female	58.0	0	0	113783	26.5500	C103	S
13	0	3	Saundercock, Mr. William Henry	male	20.0	0	0	A/5. 2151	8.0500	NaN	S
14	0	3	Andersson, Mr. Anders Johan	male	39.0	1	5	347082	31.2750	NaN	S
15	0	3	Vestrom, Miss. Hulda Amanda Adolfina	female	14.0	0	0	350406	7.8542	NaN	S

Variable	Description	Details
survival	Survival	0 = No; 1 = Yes
pclass	Passenger Class	1 = 1st; 2 = 2nd; 3 = 3rd
name	First and Last Name	
sex	Sex	
age	Age	
sibsp	Number of Siblings/Spouses Aboard	
parch	Number of Parents/Children Aboard	
ticket	Ticket Number	
fare	Passenger Fare	
cabin	Cabin	
embarked	Port of Embarkation	C = Cherbourg; Q = Queenstown; S = Southampton

Figure 4.4 – Displaying the dataset as a pandas DataFrame (left-hand side) and a data dictionary (right-hand side)

For this particular example, we are not concerned about getting a highly efficient machine learning model, but rather our focus is on using LIME to produce human-friendly explanations in a few lines of code. So, we will skip doing rigorous **Exploratory Data Analysis (EDA)** or feature engineering steps. However, I do highly encourage all of you to perform these steps as a good practice. As we can see from the dataset, certain features such as *Passenger ID* and *Ticket Number* are identifiers that can be ignored. The *Cabin Number* feature is an interesting feature, especially as it could indicate a certain wing, floor, or side of the ship that is more vulnerable. But this feature is a sparse categorical feature, which, alone, will not be very helpful and might require some advanced transformation or feature engineering. So, to build a simple model, we will drop this feature. Also, the *passenger names* are not useful for the predictive model, and hence, we can remove them. There are some categorical features that need to be transformed for better model results. If you want to try out some more ideas for feature engineering, the following article might be helpful: `https://triangleinequality.wordpress.com/2013/09/08/basic-feature-engineering-with-the-titanic-data/`.

Here are the lines of code for data preparation before the model training:

```
# Dropping all irrelevant columns
data.drop(columns=['PassengerId', 'Name', 'Cabin', 'Ticket'],
inplace = True)
# Handling missing values
data.dropna(inplace=True)
# Ensuring that Age and Fare is of type float
data['Age'] = data['Age'].astype('float')
data['Fare'] = data['Fare'].astype('float')
# Label Encoding features
categorical_feat = ['Sex']
# Using label encoder to transform string categories to integer
labels
le = LabelEncoder()
for feat in categorical_feat:
    data[feat] = le.fit_transform(data[feat]).astype('int')


# One-Hot Encoding Categorical features
data = pd.get_dummies(data, columns=['Embarked'])
```

The transformed DataFrame is shown in *Figure 4.5*:

Survived	Pclass	Sex	Age	SibSp	Parch	Fare	Embarked_0	Embarked_C	Embarked_Q	Embarked_S
0	3	1	22.0	1	0	7.2500	0	0	0	1
1	1	0	38.0	1	0	71.2833	0	1	0	0
1	3	0	26.0	0	0	7.9250	0	0	0	1
1	1	0	35.0	1	0	53.1000	0	0	0	1
0	3	1	35.0	0	0	8.0500	0	0	0	1

Figure 4.5 – DataFrame display after basic preprocessing and feature engineering

Now, for the model training part, we will use an XGBoost classifier. This is an ensemble learning algorithm and is not inherently interpretable. Based on the number of estimators, the complexity of the algorithm can vary. It can also be installed easily using the pip installer:

```
!pip install xgboost
```

The code to train the model after dividing into training and testing is as follows:

```
features = data.drop(columns=['Survived'])
labels = data['Survived']
# Dividing into training-test set with 80:20 split ratio
x_train,x_test,y_train,y_test = train_test_split(
    features,labels,test_size=0.2, random_state=123)
model = XGBClassifier(n_estimators = 300,
                      random_state = 123)
model.fit(x_train, y_train)
```

Next, let's define *f* as the prediction probability score, which will be later utilized by the LIME framework:

```
predict_fn = lambda x: model.predict_proba(x)
```

To provide model explanations, we can define the LIME object and explain the required data instance with just a few lines of code:

```
explainer = lime.lime_tabular.LimeTabularExplainer(
    data[features.columns].astype(int).values,
    mode='classification',
    training_labels=data['Survived'],
    feature_names=features.columns)
exp = explainer.explain_instance(
    data.loc[i,features.columns].astype(int).values,
    predict_fn, num_features=5)
exp.show_in_notebook(show_table=True)
```

The following diagram shows the visualizations provided by LIME for model explainability:

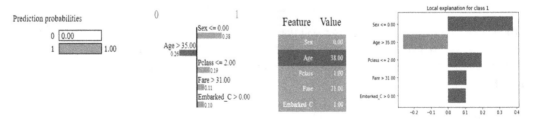

Figure 4.6 – Visualizations provided by the LIME framework to explain the model outcome

From *Figure 4.6*, we can see the explanations provided by the LIME framework with only a few lines of code. Now, let's try to understand what the visualization is telling us:

- The leftmost bar plot is showing us the prediction probabilities, which can be treated as the model's confidence level in making the prediction. In *Figure 4.6*, for the selected data instance, the model is 100% confident that the particular passenger would *survive*.

- The second visualization from the left is probably the most important visualization that provides maximum explainability. It tells us that the most important feature, with a feature importance score of 38%, is the `Sex` feature, followed by `Age`, with a feature importance score of 26%. However, as illustrated in *Figure 4.6*, for the selected data instance, the `Sex`, `Pclass` (Passenger Class), `Fare`, and `Embarked_C` (Port of Embarkation as Cherbourg) features contribute toward the model outcome of *survival* along with their threshold scores learned from the entire dataset. In comparison, the `Age` feature, which is highlighted in blue, was more inclined toward predicting the outcome as *Did not Survive* as the particular passenger's age was 38 and, usually, passengers above the age of 38 have lower chances of surviving the disaster. The threshold feature values learned by the LIME model are also in alignment with our own common sense and *a prior* knowledge. Even in the case of the actual incident of the sinking of the Titanic, which happened over 100 years ago, women and children were given the first preference to escape the sinking ship using the limited lifeboats.

  Similarly, first-class passengers who had paid higher ticket fares got a higher preference to take the lifeboats and, therefore, had higher chances of survival. So, the model explanation provided is human-friendly and consistent with our prior beliefs.

- The third visualization from the left shows the top five features and their respective values. Here, the features highlighted in orange are contributing toward class 1, while features highlighted in blue are contributing toward class 0.

- The rightmost visualization is almost the same as the second visualization, except that it is presented in a different format, and it also provides local explanations for the particular data instance selected.

As we discussed in the previous section, LIME also provides a global understanding of the model alongside the local explanations. This is provided using the SP-LIME algorithm. This can be implemented using the following lines of code:

```
sp _ exp = submodular _ pick.SubmodularPick(
    explainer, data[features.columns].values, predict _ fn,
    num _ features=5, num _ exps _ desired=10)
[exp.as _ pyplot _ figure(label=exp.available _ labels()[0]) for exp
in sp _ exp.sp _ explanations]
```

Figure 4.7 shows the visualizations obtained using SP-LIME:

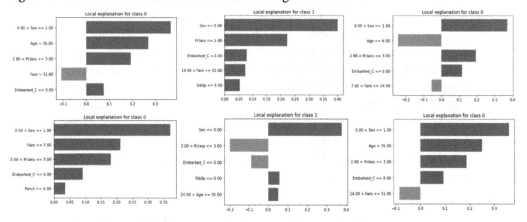

Figure 4.7 – Visualizations of diverse explanations obtained from SP-LIME
to get a global understanding of the model

Figure 4.7 shows the output of the SP-LIME code. SP-LIME provides a diverse representative sample set of local explanations considering different instances of the model to get a global perspective of the black-box model. These visualizations show us the important features, the feature-important scores, and even the range of values for each of those features and how these features contribute toward either of the classes. All these properties and features of the entire LIME framework make it a powerful approach in which to provide model-agnostic human-understandable model interpretability to black-box models. Additionally, the framework is also very robust so the entire algorithm can be implemented with only a few lines of code.

Although LIME has many advantages, unfortunately, there are some drawbacks of this algorithm that we should be aware of. Let's discuss them in the next section.

Potential pitfalls

In the previous section, we learned how easily the LIME Python framework can be used to explain black-box models for a classification problem. But unfortunately, the algorithm does have certain limitations, and there are a few scenarios in which the algorithm is not effective:

- While providing interpretable explanations, a particular choice of interpretable data representation and interpretable model might still have a lot of limitations. While the underlying trained model might still be considered a black-box model, there is no assumption about the model that is made during the explanation process. However, certain representations are not powerful enough to represent some complex behaviors of the model. For example, if we are trying to build an image classifier to distinguish between black and white images and colored images, then the presence or absence of superpixels will not be useful to provide the explanations.

- As discussed earlier, LIME learns an interpretable model to provide local explanations. Usually, these interpretable models are linear and non-complex. However, suppose that the underlying black-box model is not linear, even in the locality of the prediction, so the LIME algorithm is not effective.

- LIME explanations are highly sensitive to any change in input data. Even a slight change in the input data can drastically alter the explanation instance provided by LIME.

- For certain datasets, LIME explanations are not robust as, even for similar data instances, the explanations provided can be completely different. This might prevent end users from completely relying on the explanations provided by LIME.

- The algorithm is extremely prone to data drifts and label drifts. A slight drift between the training and the inference data can completely produce inconsistent explanations. The authors of the paper named *A study of data and label shift in the LIME framework, Rahnama* and *Boström* (https://arxiv.org/abs/1910.14421), mention certain experiments that can be used to evaluate the impact of data drift in the LIME framework. Due to this limitation, the goodness of approximation of the LIME explanations (also referred to as *fidelity*) is considered to be low. This is not expected in a good explanation method.

- Explanations provided by LIME depend on the choice of the hyperparameters of the algorithm. Similar to most of the algorithms, even for the LIME algorithm, the choice of the hyperparameters can determine the quality of the explanations provided. Hyperparameter tuning is also difficult for the LIME algorithm as, usually, qualitative methods are adopted to evaluate the quality of the LIME explanations.

There are many research works that indicate the other limitations of the LIME algorithm. I have mentioned some of these research works in the *References* section. I would strongly recommend that you go through those papers to get more details about certain limitations of the algorithm.

Summary

This brings us to the end of the chapter. In this chapter, we discussed LIME, one of the most widely adopted frameworks in XAI. Throughout this chapter, we discussed the intuition behind the workings of the algorithm and some important properties of the algorithm that make the generated explanations human-friendly. Additionally, we saw an end-to-end tutorial on how to use LIME for a practical use case to provide explainability to a black-box classification model. Even though we discussed some limitations of the LIME algorithm, due to its simplicity, LIME is still one of the most popular and widely used XAI frameworks. Hence, it is very important for us to discuss this algorithm and have a thorough understanding of the workings of the framework.

In the next chapter, we will apply the LIME framework to solve other types of machine learning problems using different types of datasets.

References

For additional information, please refer to the following resources:

- *"Why Should I Trust You?" Explaining the Predictions of Any Classifier* by *Ribeiro et al*: `https://arxiv.org/pdf/1602.04938.pdf`

- *LIME - Local Interpretable Model-Agnostic Explanations*: `https://homes.cs.washington.edu/~marcotcr/blog/lime/`

- The LIME GitHub project: `https://github.com/marcotcr/lime`

- *A study of data and label shift in the LIME framework* by *Rahnama* and *Boström*: `https://arxiv.org/abs/1910.14421`

- *What's Wrong with LIME*: `https://towardsdatascience.com/whats-wrong-with-lime-86b335f34612`

- *Why model why? Assessing the strengths and limitations of LIME* by *Dieber* and *Kirrane* (2020): `https://arxiv.org/pdf/2012.00093.pdf`

5
Practical Exposure to Using LIME in ML

After reading the last chapter, you should now have a good conceptual understanding of **Local Interpretable Model-agnostic Explanations (LIME)**. We saw how the LIME Python framework can explain black-box models for classification problems. We also discussed some of the pros and cons of the LIME framework. In practice, LIME is still one of the most popular XAI frameworks as it can be easily applied to tabular datasets and text and image datasets. LIME can provide model-agnostic local explanations for solving both regression and classification problems.

In this chapter, you will get much more in-depth practical exposure to using LIME in ML. These are the main topics of discussion for this chapter:

- Using LIME on tabular data
- Explaining image classifiers with LIME
- Using LIME on text data
- LIME for production-level systems

Technical requirements

Like the previous chapter, this chapter is very technical with code walk-throughs in Python and Jupyter notebooks. For this chapter, the code and dataset resources can be downloaded or cloned from the GitHub repository: `https://github.com/PacktPublishing/Applied-Machine-Learning-Explainability-Techniques/tree/main/Chapter05`. Like the previous chapters, we will be using Python and Jupyter notebooks to run the code and generate the necessary outputs. All other relevant details are provided in the notebooks, and I recommend that you all run the notebooks while going through the chapter content to get a better understanding of the topics covered.

Using LIME on tabular data

In the *Practical example of using LIME for classification problems* section of *Chapter 4, LIME for Model Interpretability*, we discussed how to set up LIME in Python and how to use LIME to explain classification ML models. The dataset used for the tutorial in *Chapter 4, LIME for Model Interpretability* (`https://github.com/PacktPublishing/Applied-Machine-Learning-Explainability-Techniques/blob/main/Chapter04/Intro_to_LIME.ipynb`) was a tabular structured data. In this section, we will discuss using LIME to explain regression models that are built on tabular data.

Setting up LIME

Before starting the code walk-through, I would ask you to check the following notebook, `https://github.com/PacktPublishing/Applied-Machine-Learning-Explainability-Techniques/blob/main/Chapter05/LIME_with_tabular_data.ipynb`, which already contains the steps needed to understand the concept that we are going to discuss now in more depth. I assume that most of the Python libraries that we will use for this tutorial are already installed on your system. But if not, please run the following command to install the upgraded versions of the Python libraries that we are going to use:

```
!pip install --upgrade pandas numpy matplotlib seaborn scikit-learn lime
```

Discussion about the dataset

For this tutorial, we will use the *Diabetes dataset* from *scikit-learn datasets* (`https://scikit-learn.org/stable/datasets/toy_dataset.html#diabetes-dataset`). This dataset is used to predict the *disease progression level* of diabetes. It contains around *442 samples* with *10 baseline features* – age, sex, body mass index (bmi), average blood pressure (bp), total serum cholesterol (s1), low-density lipoproteins (s2), high-density lipoproteins (s3), total cholesterol / HDL (s4), possibly log of serum triglycerides level (s5), and blood sugar level (s6). The dataset is quite interesting and relevant, considering that the underlying problem of monitoring diabetes progression is an important practical problem.

The feature variables provided in the dataset are already normalized by centering the feature values around the mean and scaling by the standard deviation times the number of samples (N):

$$normalized_x = \frac{x - mean(x)}{standard_{deviation(x)} * N}$$

More information about the original dataset can be found at `https://www4.stat.ncsu.edu/~boos/var.select/diabetes.html`.

To load the dataset, just execute the following lines of code:

```
from sklearn import datasets
dataset = datasets.load_diabetes()
features, labels = dataset.data, dataset.target
```

You can perform the necessary EDA steps if needed, but since our main focus is to use LIME for explaining black-box models, we will not spend too much effort on EDA for the purpose of this tutorial.

Discussions about the model

As demonstrated in the notebook tutorial, we have used a **Gradient Boosting Regressor (GBR)** (`https://scikit-learn.org/stable/modules/generated/sklearn.ensemble.GradientBoostingRegressor.html`) model to train our predictive model. However, any regression ML algorithm can be used instead of GBR as the model itself is regarded as any black-box model by the LIME algorithm. Also, when we evaluated the trained model on the unseen data, we observed a **Mean Absolute Percentage Error (MAPE)** of 0.37, a **Mean Square Error (MSE)** of 2,538, and an **R-squared coefficient** score of 0.6. All these results indicate that our model is not very good and definitely has room for improvement. So, if such a model is deployed in production-level systems, there can be many questions asked by the end stakeholders as it is always difficult for them to trust models that are not accurate. Also, algorithms such as GBR are not inherently interpretable and the complexity of the algorithm depends on hyperparameters including the number of estimators and the depth of the tree. Thus, model explainability frameworks such as LIME are not just an add-on step, but a necessary part of the process of building ML models. Next, we will see how easily LIME can be applied to explain black-box regression models with just a few lines of code.

Application of LIME

As we have seen in the previous chapter, we can easily support the LIME framework for tabular data with the following commands:

```
import lime
import lime.lime_tabular
```

Once the LIME module is successfully imported, we will need to create an explainer object:

```
explainer = lime.lime_tabular.LimeTabularExplainer(
    x_train, mode='regression',
    class_names=['disease_progression'],
    feature_names=dataset.feature_names)
```

Then, we just need to take the data instance and provide local explainability to it:

```
exp = explainer.explain_instance(x_test[i], model.predict,
                                 num_features=5)
exp.show_in_notebook(show_table=True)
```

We will get the following output from the preceding lines of code:

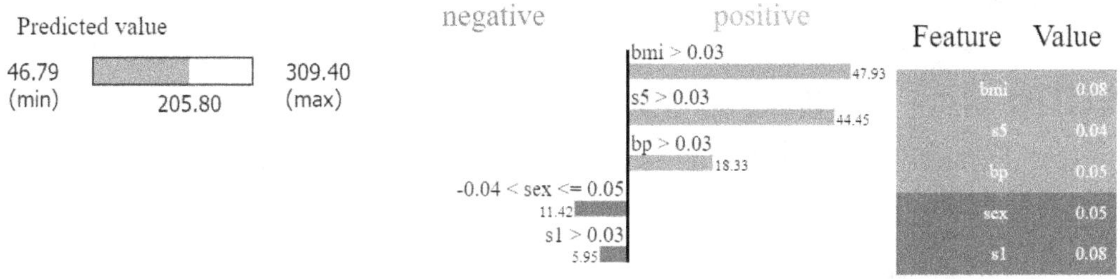

Figure 5.1 – Output visualization from the LIME framework when applied
to a regression model trained on a tabular dataset

Figure 5.1 illustrates the visualization-based explanations provided by the LIME
framework.

Next, let's try to understand what the output visualization in *Figure 5.1* is telling us:

- The left-most visualization from *Figure 5.1* shows a range of possible values and
  the position of the model's predicted outcome. Intuitively speaking, all model
  predictions should lie within the minimum and the maximum possible value as
  this indicates to the user to compare the current forecast with the best-case and the
  worst-case values.

- The middle visualization shows which features contribute to the prediction being
  on the higher side or the lower side. Considering our prior knowledge of diabetes,
  a higher BMI, as well as raised blood pressure and serum triglyceride levels, do
  indicate increasing progression of the disease.

- The right-most visualization in *Figure 5.1* shows us the actual local data values
  for the most important features identified, arranged in descending order of their
  relevance.

The explanations provided by the LIME framework are human-interpretable to a great
extent and do give us an indication of the feature-value pairs used by the black-box model
to make predictions.

So, this is how we can use LIME to explain black-box regression models trained on tabular
data with just a few lines of code. But, as we discussed in *Chapter 4, LIME for Model
Interpretability*, under the *Potential pitfalls* section, explanations provided by LIME are
not always holistic and may have some inconsistencies. This is something we all need to
be mindful of. However, LIME explanations, coupled with a thorough EDA, data-centric
XAI, counterfactual explanations, and other model explainability methods, can provide
a powerful, holistic explainability to black-box models trained on tabular datasets.

Now, let's explore how to use LIME for classifiers trained on unstructured data such as images in the next section.

Explaining image classifiers with LIME

In the previous section, we have seen how we can easily apply LIME to explain models trained on tabular data. However, the main challenge always comes while explaining complex deep learning models trained on unstructured data such as images. Generally, deep learning models are much more efficient than conventional ML models on image data as these models have the ability to perform *auto feature extraction*. They can extract complex *low-level features* such as *stripes*, *edges*, *contours*, *corners*, and *motifs*, and even *higher-level features* such as *larger shapes* and *certain parts of the object*. These higher-level features are usually referred to as **Regions of Interest (RoI)** in the image, or **superpixels**, as they are collections of pixels of the image that cover a particular area of the image. Now, the low-level features are not human-interpretable, but the high-level features are human-interpretable, as any non-technical end user will relate to the images with respect to the higher-level features. LIME also works in a similar fashion. The algorithm tries to highlight the superpixels in images that contribute positively or negatively to the model's decision-making process. So, let's see how LIME can be used to explain image classifiers.

Setting up the required Python modules

Before we begin the code walk-through, please check the notebook provided in the code repository: `https://github.com/PacktPublishing/Applied-Machine-Learning-Explainability-Techniques/blob/main/Chapter05/LIME_with_image_data.ipynb`. The notebook contains the necessary details required for the practical application of the concepts. In this section, I will give you a walk-through of the code and explain all the steps covered in the notebook tutorial. Use the following command to install the upgraded versions of the Python libraries if not already installed:

```
!pip install --upgrade pandas numpy matplotlib seaborn
tensorflow lime scikit-image
```

Next, let's discuss the model used in this example.

Using a pre-trained TensorFlow model as our black-box model

For this tutorial, we have used a *pre-trained TensorFlow Keras Xception model* as our black-box model. The model is pre-trained on the ImageNet dataset (https://www.image-net.org/), which is one of the most popular benchmarking datasets for image classification. The pre-trained model can be loaded with the following lines of code:

```
from tensorflow.keras.applications.xception import Xception
model = Xception(weights="imagenet")
```

In order to use any inference data for image classification, we will also need to perform the necessary preprocessing steps. Please refer to the notebook at https://github.com/PacktPublishing/Applied-Machine-Learning-Explainability-Techniques/blob/main/Chapter05/LIME_with_image_data.ipynb for the necessary pre-processing methods.

Application of LIME Image Explainers

In this subsection, we will see how the LIME framework can be used to identify *super-pixels* or regions from the image used by the model to predict the specific outcome. We will first need to define an image explainer object:

```
explainer = lime_image.LimeImageExplainer()
```

Next, we will need to pass the inference data (normalized_img[0]) to the explainer object and use the LIME framework to highlight superpixels that have the maximum positive and negative influence on the model's prediction:

```
exp = explainer.explain_instance(normalized_img[0],
                                 model.predict,
                                 top_labels=5,
                                 hide_color=0,
                                 num_samples=1000)


image, mask = exp.get_image_and_mask(exp_class,
                                     positive_only=False,
                                     num_features=6,
                                     hide_rest=False,
                                     min_weight=0.01)
```

```
plt.imshow(mark_boundaries(image, mask))
plt.axis('off')
plt.show()
```

As an output to the preceding lines of code, we will get certain highlighted portions of the image that contribute to the model's prediction, in both a positive and negative manner:

Figure 5.2 – (Left) Original inference image. (Middle) Most important image superpixel. (Right) Image with superposed mask superpixel on the original data highlighted in green

In *Figure 5.2*, the left-most image was used as the inference image. When the trained model was applied to the inference image, the top prediction of the model was a *tiger shark*.

The prediction was actually correct. However, in order to explain the model, the LIME algorithm can highlight the superpixel, which has the maximum influence on the prediction. From the middle and the right-most images in *Figure 5.2*, we can see that the black-box model was actually good and trustworthy as the relevant superpixel captured by the LIME algorithm indicates the presence of a tiger shark.

The superpixel estimated by the LIME algorithm can be displayed using the following lines of code:

```
plt.imshow(exp.segments)
plt.axis('off')
plt.show()
```

We can also form a heatmap highlighting the importance of each of the superpixels, which gives us further insight into the functioning of the black-box model:

```
dict_heatmap = dict(exp.local_exp[exp.top_labels[0]])
heatmap = np.vectorize(dict_heatmap.get)(exp.segments)
plt.imshow(heatmap, cmap = 'RdBu', vmin  = -heatmap.max(),
```

```
                vmax = heatmap.max())
plt.colorbar()
plt.show()
```

The output obtained is shown in *Figure 5.3*:

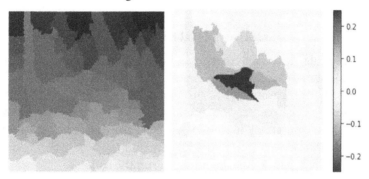

Figure 5.3 – (Left) Image showing all the superpixels picked up by the LIME algorithm.
(Right) Heatmap of the superpixels based on their importance in terms of the model's prediction

The heatmap from *Figure 5.3* provides us with some insight into important superpixels, which is also easy when it comes to any non-technical user interpreting any black-box model.

So, we have seen how LIME can explain even complicated deep learning models trained on image data in just a few lines of code. I found LIME to be one of the most effective algorithms to visually explain deep learning-based image classifiers without presenting any complicated statistical or numerical values or complicated graphical visualizations. Unlike tabular data, I felt explanations provided to image classifiers are more robust, stable, and human-interpretable. It is definitely one of my favorite methods for interpreting image classifiers, and before moving any image classification model to production, I strongly recommend applying LIME as an additional evaluation step to gain more confidence in the trained model.

In the next section, let's explore LIME for models trained on text data.

Using LIME on text data

In the previous section, we discussed how LIME is an effective approach to explaining complicated black-box models trained on image datasets. Like images, text is also a form of unstructured data, which is very much different from structured tabular data. Explaining such black-box models trained on unstructured data is always very challenging. But LIME can also be applied to models trained on text data.

Using the LIME algorithm, we can analyze whether the presence of a particular word or group of words increases the probability of predicting a specific outcome. In other words, LIME helps to highlight the importance of text tokens or words that can influence the model's outcome toward a particular class. In this section, we will see how LIME can be used to interpret text classifiers.

Installing the required Python modules

Like the previous tutorials, the complete notebook tutorial is available at `https://github.com/PacktPublishing/Applied-Machine-Learning-Explainability-Techniques/blob/main/Chapter05/LIME_with_text_data.ipynb`. Although the necessary instructions needed to run the notebook are clearly documented in the notebook itself, similar to the previous tutorials, I will provide the necessary details to walk you through the implementation. Using the following commands, you can install the modules required to run the code:

```
!pip install --upgrade pandas numpy matplotlib seaborn scikit-learn nltk lime xgboost swifter
```

For text-related operations on the underlying dataset, I will be mainly using the NLTK Python framework. So, you will need to download certain `nltk` modules by executing the following commands:

```
nltk.download('stopwords')
nltk.download('wordnet')
nltk.download('punkt')
nltk.download('averaged_perceptron_tagger')
```

In the tutorial, we will try to explain a text classifier designed to perform sentiment analysis by classifying the text data into positive and negative classes.

Discussions about the dataset used for training the model

We have used the **Sentiment Polarity dataset v2.0** consisting of *Movie reviews* for this tutorial used for sentiment analysis from text data. The dataset consists of about 1,000 samples of positive and negative movie reviews. More information about the dataset can be found on the source website: `https://www.cs.cornell.edu/people/pabo/movie-review-data/`. The dataset is also provided in the GitHub repository of this chapter: `https://github.com/PacktPublishing/Applied-Machine-Learning-Explainability-Techniques/tree/main/Chapter05`.

> **Sentiment Polarity dataset v2.0**
>
> This data was first used in Bo Pang and Lillian Lee, "A Sentimental Education: Sentiment Analysis Using Subjectivity Summarization Based on Minimum Cuts", Proceedings of the ACL, 2004.

Discussions about the text classification model

Unlike the previous image classifier tutorial, we have not used a pre-trained model. We have trained an **XGBoost Classifier** (`https://xgboost.readthedocs.io/en/stable/`) from scratch with the necessary data preprocessing, preparation, and feature extraction steps as covered in the notebook. XGBoost is an ensemble learning boosting algorithm that is not inherently interpretable. So, we will consider this as our black-box text classification model. We are not focused on improving the model's accuracy with necessary hyperparameter tuning, as LIME is completely model-agnostic. For this tutorial, we have created a scikit-learn pipeline to apply feature extraction first using **TFIDF Vectorizer** (`https://scikit-learn.org/stable/modules/generated/sklearn.feature_extraction.text.TfidfVectorizer.html`), and then the trained model:

```
model_pipeline = make_pipeline(tfidf, model)
```

In the next subsection, we will see how the LIME framework can be easily applied with text data as well.

Applying LIME Text Explainers

Like the previous tutorials with image and tabular data, applying LIME is simple with text data in a few lines of code. We will now define the LIME `explainer` object:

```
from lime.lime_text import LimeTextExplainer
explainer = LimeTextExplainer(class_names=['negative',
'positive'])
```

Then, we will use the inference data instance to provide local explainability for that particular data instance:

```
exp = explainer.explain_instance(x_test_df[idx], model_
pipeline.predict_proba, num_features=5)
exp.show_in_notebook()
```

And that's it! In just a few lines of code, we can explain the text classifier that actually relies on TFIDF numerical features, but the explainability is provided with a human interpretable perspective as words that can positively or negatively influence the model outcome are highlighted. It is easier for any non-technical user to understand the working of the text model in this way, rather than providing explanations using numerically encoded features.

Now, let's take a look at the output visualization provided by LIME when applied to text data.

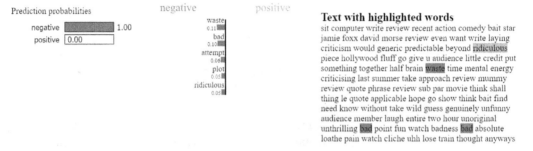

Figure 5.4 – Output visualization when LIME is applied to a text classifier

In *Figure 5.4*, we can see the output visualization of the LIME framework when applied to text data.

The output visualization is very similar to what we have observed with tabular data. It shows us the *prediction probability*, which can be used as a *model confidence* score. The algorithm highlights the most influential words that determine the model outcome, with a feature importance score. For example, from *Figure 5.4*, we can see that the inference data instance is predicted as negative by the model (which is predicted correctly as demonstrated in the notebook). The presence of words such as *waste*, *bad*, and *ridiculous* does indicate a *negative review*. This is human-interpretable as well, since if you ask a non-technical user to justify why the review is classified as negative, the user might refer to the usage of frequently used words in negative reviews or words used in sentences with a negative tone.

Thus, we can see that LIME can be easily applied with text classifiers as well. Even with text data, the algorithm is simple, yet effective in providing a human-interpretable explanation. I would definitely recommend using LIME to explain black-box text classifiers as an additional model evaluation or quality inspection step.

But so far, we have seen the application of LIME Python frameworks in the Jupyter notebook environment. The immediate question that you might have is – *Can we scale LIME for use in production-level systems?* Let's find out in the next section.

LIME for production-level systems

The short answer to the question posted toward the end of the last section is *Yes*. LIME can definitely be scaled for use in production-level systems due to the following main reasons:

- **Minimal implementation complexity**: The API structure of the LIME Python framework is concise and well structured. This allows us to add model explainability in just a few lines of code. For providing local explainability to inference data instances, the runtime complexity of the LIME algorithm is very low and, hence, this approach can also work for real-time applications.

- **Easy integration with other software applications**: The API structure of the framework is modular. For consuming the explainability results, we do not need to solely depend on the in-built visualizations provided by the framework. We can utilize the raw explainability results and create our own custom visualization dashboards or reports. Also, we can create custom web API methods and host the web APIs on remote cloud servers, creating our own model explainability cloud-based service that can be integrated easily with other software applications. We will cover this in more detail in *Chapter 10, XAI Industry Best Practices*.

- **Does not require heavy computational resources**: The LIME framework works well with low computational resources. For real-time applications, the algorithms used need to be very fast and should have the ability to run on low computational resources, as otherwise, the user experience is affected.

- **Easy to set up and package**: As we have already seen before running the tutorial notebooks, LIME is very easy to set up and does not have a dependency on packages that are difficult to install. Similarly, any Python program using LIME is easy to package or **containerize**. Most production-level systems have automated CI/CD pipelines to create **Docker containers** (`https://www.docker.com/resources/what-container`), which are deployed on production-level systems. The engineering effort needed to containerize a Python program using a LIME framework is low and hence it is easy to productionalize such software applications.

These are the key reasons why LIME is the preferred model explainability method used in industrial applications, despite some of its well-known pitfalls.

Summary

In this chapter, we have discussed the practical applications of the LIME Python framework on different types of datasets. The tutorials covered in the chapter are just the starting point and I strongly recommend you try out LIME explainability on other datasets. We have also discussed why LIME is a good fit for production-level ML systems.

In the next chapter, we will discuss another very popular explainable AI Python framework called **SHAP**, which even considers the collective contribution of multiple features in influencing the model's outcome.

References

Please refer to the following resources to gain additional information:

- *"Why Should I Trust You?" Explaining the Predictions of Any Classifier*, by *Ribeiro et al.*: https://arxiv.org/pdf/1602.04938.pdf

- *Local Interpretable Model-Agnostic Explanations (LIME): An Introduction*: https://www.oreilly.com/content/introduction-to-local-interpretable-model-agnostic-explanations-lime/

- *LIME GitHub Project*: https://github.com/marcotcr/lime

- *Docker Blog*: https://www.docker.com/blog/

6
Model Interpretability Using SHAP

In the previous two chapters, we explored model-agnostic local explainability using the **LIME** framework to explain black-box models. We also discussed certain limitations of the LIME approach, even though it remains one of the most popular **Explainable AI (XAI)** algorithms. In this chapter, we will cover **SHapley Additive exPlanation (SHAP)**, which is another popular XAI framework that can provide model-agnostic local explainability for tabular, image, and text datasets.

SHAP is based on **Shapley values**, which is a concept popularly used in **Game Theory** (https://c3.ai/glossary/data-science/shapley-values/). Although the mathematical understanding of Shapley values can be complicated, I will provide a simple, intuitive understanding of Shapley values and SHAP and focus more on the practical aspects of the framework. Similar to LIME, SHAP also has its pros and cons, which we are going to discuss in this chapter. This chapter will cover one practical tutorial that will explain regression models using SHAP. Later, in *Chapter 7, Practical Exposure to Using SHAP in ML*, we will cover other practical applications of the SHAP framework.

So, here is a list of the main topics of discussion for this chapter:

- An intuitive understanding of the SHAP and Shapley values
- Model explainability approaches using SHAP
- Advantages and limitations
- Using SHAP to explain regression models

Now, let's begin!

Technical requirements

The code tutorial with the necessary resources can be downloaded or cloned from the GitHub repository for this chapter: `https://github.com/PacktPublishing/Applied-Machine-Learning-Explainability-Techniques/tree/main/Chapter06`. Python and Jupyter notebooks are used to implement the practical application of the theoretical concepts that are covered in this chapter. But I recommend that you only run the notebooks after you go through this chapter for a better understanding. Additionally, please look at the *SHAP Errata* section before proceeding with the practical tutorial part of this chapter: `https://github.com/PacktPublishing/Applied-Machine-Learning-Explainability-Techniques/blob/main/Chapter06/SHAP_ERRATA/ReadMe.md`.

An intuitive understanding of the SHAP and Shapley values

As discussed in *Chapter 1*, *Foundational Concepts of Explainability Techniques*, explaining black-box models is a necessity for increasing AI adoption. Algorithms that are model-agnostic and can provide local explainability with a global perspective are the ideal choice of explainability technique in **machine learning (ML)** . That is why LIME is a popular choice in XAI. SHAP is another popular choice of explainability technique in ML and, in certain scenarios, is more effective than LIME. In this section, we will discuss about the intuitive understanding of the SHAP framework along with how it functions for providing model explainability.

Introduction to SHAP and Shapley values

The SHAP framework was introduced by *Scott Lundberg* and *Su-In Lee* in their research work, *A Unified Approach of Interpreting Model Predictions* (`https://arxiv.org/abs/1705.07874`). This was published in 2017. SHAP is based on the concept of Shapley values from cooperative game theory, but unlike the LIME framework, it considers *additive feature importance*. By definition, the Shapley value is *the mean marginal contribution of each feature value across all possible values in the feature space*. The mathematical understanding of Shapley values is complicated and might confuse most readers. That said, if you are interested in getting an in-depth mathematical understanding of Shapley values, we recommend that you take a look at the research paper called *"A Value for n-Person Games." Contributions to the Theory of Games 2.28 (1953), by Lloyd S. Shapley*. In the next section, we will gain an intuitive understanding of Shapley values with a very simple example.

What are Shapley values?

In this section, I will explain Shapley values using a very simple and easy-to-understand example. Let's suppose that Alice, Bob, and Charlie are three friends who are taking part, as a team, in a Kaggle competition to solve a given problem with ML, for a certain cash prize. Their collective goal is to win the competition and get the prize money. All three of them are equally not good in all areas of ML and, therefore, have contributed in different ways. Now, if they win the competition and earn their prize money, *how will they ensure a fair distribution of the prize money considering their individual contributions? How will they measure their individual contributions for the same goal?* The answer to these questions can be given by Shapley values, which were introduced in 1951 by Lloyd Shapley.

The following diagram gives us a visual illustration of the scenario:

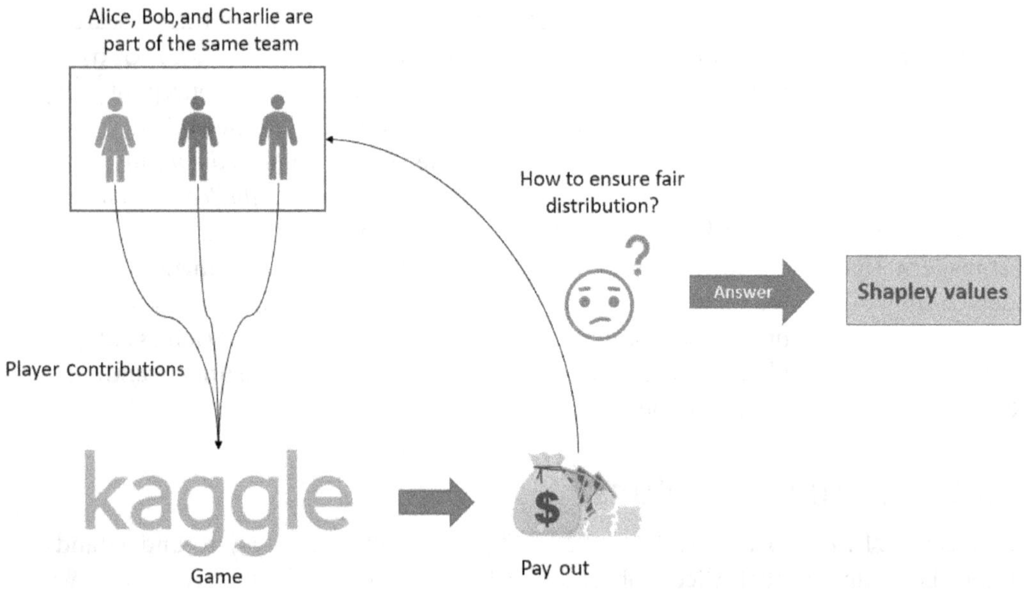

Figure 6.1 – Visual illustration of the scenario discussed in the What are Shapley values? section

So, in this scenario, Alice, Bob, and Charlie are part of the same team, playing the same game (which is the Kaggle competition). In game theory, this is referred to as a **Coalition Game**. The prize money for the competition is their *payout*. So, Shapley values tell us the average contribution of each player to the payout ensuring a fair distribution. But *why not just equally distribute the prize money between all the players*? Well, since the contributions are not equal, it is not *fair* to distribute the money equally.

Deciding the payouts

Now, how do we decide the fairest way to distribute the payout? One way is to assume that Alice, Bob, and Charlie joined the game in a sequence in which Alice started first, followed by Bob, and then followed by Charlie. Let's suppose that if Alice, Bob, and Charlie had participated alone, they would have gained 10 points, 20 points, and 25 points, respectively. But if only Alice and Bob teamed up, they might have received 40 points. While Alice and Charlie together could get 30 points, Bob and Charlie together could get 50 points. When all three of them collaborate together, only then do they get 90 points, which is sufficient for them to win the competition.

Figure 6.2 illustrates the point values for each condition. We will make use of these values to calculate the average marginal contribution of each player:

Players	Point Values = V (Players)
Alice	10
Bob	20
Charlie	25
Alice, Bob	40
Alice, Charlie	30
Bob, Charlie	50
Alice, Bob, Charlie	90

Figure 6.2 – The contribution values for all possible combinations of all the players

Mathematically, if we assume that there are N players, where S is the coalition subset of players and $v(S)$ is the total value of S players, then by the formula of Shapley values, the marginal contribution of player i is given as follows:

$$\varphi(i) = \sum_{S \subseteq N/i} \frac{|S|!\,(|N| - |S| - 1)!}{|N|!} \left(v(S \cup \{i\}) - v(S) \right)$$

The equation of Shapley value might look complicated, but let's simplify this with our example. Please note that the order in which each player starts the game is important to consider as Shapley values try to account for the order of each player to calculate the marginal contribution.

Now, for our example, the contribution of Alice can be calculated by calculating the difference that Alice can cause to the final score. So, the contribution is calculated by taking the difference in the points scored when Alice is in the game and when she is not. Also, when Alice is playing, she can either play alone or team up with others. When Alice is playing, the value that she can create can be represented as $v(A)$. Likewise, $v(B)$ and $v(C)$ denote individual values created by Bob and Charlie. Now, when Alice and Bob are teaming up, we can calculate only Alice's contribution by removing Bob's contribution from the overall contribution. This can be represented as $v(A, B) - v(B)$. And if all three are playing together, Alice's contribution is given as $v(A, B, C) - v(B, C)$.

Considering all possible permutations of the sequences by which Alice, Bob, and Charlie play the game, the marginal contribution of Alice is the average of her individual contributions in all possible scenarios. This is illustrated in *Figure 6.3*:

Order	Situation	Contribution of Alice
Alice, Bob, Charlie	Alice plays alone	$v(A) - v(\varphi) = 10 - 0 = 10$
Alice, Charlie, Bob	Alice plays alone	$v(A) - v(\varphi) = 10 - 0 = 10$
Bob, **Alice**, Charlie	Alice teams with only Bob	$v(A,B) - v(B) = 40 - 20 = 20$
Charlie, **Alice**, Bob	Alice teams with only Charlie	$v(A,C) - v(C) = 30 - 25 = 5$
Bob, Charlie, **Alice**	Alice teams with both Bob and Charlie	$v(A, B, C) - v(B,C) = 90 - 50 = 40$
Charlie, Bob, **Alice**	Alice teams with both Bob and Charlie	$v(A, B, C) - v(C,B) = 90 - 50 = 40$
Shapley Value of Alice		$(10+10+20+5+40+40)/6 = $ **20.83**

Figure 6.3 – The Shapley value of Alice is her marginal contribution considering all possible scenarios

So, the overall contribution of Alice will be her marginal contribution across all possible scenarios, which also happens to be the Shapley value. For Alice, the Shapley value is *20.83*. Similarly, we can calculate the marginal contribution for Bob and Charlie, as shown in *Figure 6.4*:

Combination	Marginal Contribution		
	Alice	Bob	Charlie
Alice, Bob, Charlie	$v(A) - v(\varphi) = 10 - 0 = 10$	$v(A,B) - v(A) = 40 - 10 = 30$	$v(A,B,C) - v(A,B) = 90 - 40 = 50$
Alice, Charlie, Bob	$v(A) - v(\varphi) = 10 - 0 = 10$	$v(A,C,B) - v(A,C) = 90 - 30 = 60$	$v(A,C) - v(A) = 30 - 10 = 20$
Bob, Alice, Charlie	$v(A,B) - v(B) = 40 - 20 = 20$	$v(B) - v(\varphi) = 20 - 0 = 20$	$v(A,B,C) - v(A,B) = 90 - 40 = 50$
Charlie, Alice, Bob	$v(A,C) - v(C) = 30 - 25 = 5$	$v(A, B, C) - v(A,C) = 90 - 30 = 60$	$v(C) - v(\varphi) = 25 - 0 = 25$
Bob, Charlie, Alice	$v(A, B, C) - v(A,B) = 90 - 50 = 40$	$v(B) - v(\varphi) = 20 - 0 = 20$	$v(B,C) - v(B) = 50 - 20 = 30$
Charlie, Bob, Alice	$v(A, B, C) - v(A,B) = 90 - 50 = 40$	$v(B, C) - v(C) = 50 - 25 = 25$	$v(C) - v(\varphi) = 25 - 0 = 25$
Shapley Values	$(10+10+20+5+40+40)/6 = $ **20.83**	$(30+60+20+20+60+25)/6 = $ **35.83**	$(40+20+40+25+30+25)/6 = $ **33.34**

Figure 6.4 – Marginal contribution for Alice, Bob, and Charlie

I hope this wasn't too difficult to understand! One thing to note is that the sum of marginal contributions of Alice, Bob, and Charlie should be equal to the total contribution made by all three of them together. Now, let's try to understand Shapley values in the context of ML.

Shapley values in ML

In order to understand the importance of Shapley values in ML to explain model predictions, we will try to modify the example about Alice, Bob, and Charlie that we used for understanding Shapley values. We can consider Alice, Bob, and Charlie to be *three different features present in a dataset used for training a model.* So, in this case, the *player contributions* will be the *contribution of each feature. The game* or the Kaggle competition will be the *black-box ML model* and the *payout* will be the *prediction*. So, if we want to know the *contribution of each feature toward the model prediction*, we will use *Shapley values*.

The modification of *Figure 6.1* to represent Shapley values in the context of ML is illustrated in *Figure 6.5*:

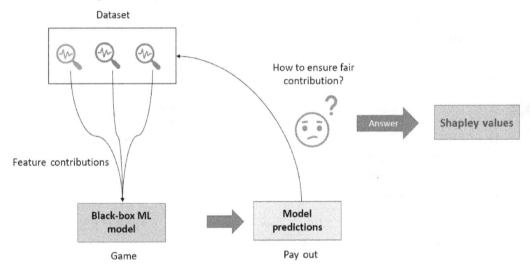

Figure 6.5 – Understanding Shapley values in the context of ML

Therefore, Shapley values help us to understand the collective contribution of each feature toward the outcome predicted by black-box ML models. By using Shapley values, we can explain the working of black-box models by estimating the feature contributions.

Properties of Shapley values

Now that we have an intuitive understanding of Shapley values and we have learned how to calculate Shapley values, we should also gain an understanding of the properties of Shapley values:

- **Efficiency**: The total sum of Shapley values or the marginal contribution of each feature should be equal to the value of the total coalition. For example, in *Figure 6.4*, we can see that sum of individual Shapley values for Alice, Bob, and Charlie are equal to the total coalition value obtained when Alice, Bob, and Charlie team up together.

- **Symmetry**: Each player has a fair chance of joining the game in any order. in *Figure 6.4*, we can see that all permutations of the sequences for all the players are considered.

- **Dummy**: If a particular feature does not change the predicted value regardless of the coalition group, then the Shapley value for the feature is *0*.

- **Additivity**: For any game with a combined payout, the Shapley values are also combined. This is denoted as $\varphi(v + w) = \varphi(v) + \varphi(w)$, then $(v + w)(S) = v(S) + w(S)$. For example, for the random forest algorithm in ML, Shapley values can be calculated for a particular feature by calculating it for each individual tree and then averaging them to find the additive Shapley value for the entire random forest.

So, these are the important properties of Shapley values. Next, let's discuss the SHAP framework and understand how it is much more than just the usage of Shapley values.

The SHAP framework

Previously, we discussed what Shapley values are and how they are used in ML. Now, let's cover the SHAP framework. Although SHAP is popularly used as an XAI tool for providing local explainability to individual predictions, SHAP can also provide a global explanation by aggregating the individual predictions. Additionally, SHAP is *model-agnostic*, which means that it does not make any assumptions about the algorithm used in black-box models. The creators of the framework broadly came up with two model-agnostic approximation methods, which are as follows:

- **SHAP Explainer**: This is based on *Shapley sampling values*.
- **KernelSHAP Explainer**: This is based on the *LIME approach*.

The framework also includes *model-specific* explainability methods such as the following:

- **Linear SHAP**: This is for linear models with independent features.

- **Tree SHAP**: This is an algorithm that is faster than SHAP explainers to compute SHAP values for tree algorithms and tree-based ensemble learning algorithms.

- **Deep SHAP**: This is an algorithm that is faster than SHAP explainers to compute SHAP values for deep learning models.

Apart from these approaches, SHAP also uses interesting visualization methods to explain AI models. We will cover these methods in more detail in the next section. But one point to note is that the calculation of Shapley values is computationally very expensive, and the algorithm is of the order of $O(2^n)$, where n is the number of features. So, if the dataset has many features, calculating Shapley values might take forever! However, the SHAP framework uses an approximation technique to calculate Shapley values efficiently. The explanation provided by SHAP is more robust as compared to the LIME framework. Let's proceed to the next section, where we will discuss the various model explainability approaches used by SHAP on various types of data.

Model explainability approaches using SHAP

After reading the previous section, you have gained an understanding of SHAP and Shapley values. In this section, we will discuss various model-explainability approaches using SHAP. Data visualization is an important method to explain the working of complex algorithms. SHAP makes use of various interesting data visualization techniques to represent the approximated Shapley values to explain black-box models. So, let's discuss some of the popular visualization methods used by the SHAP framework.

Visualizations in SHAP

As mentioned previously, SHAP can be used for both the global interpretability of the model and the local interpretability of the inference data instance. Now, the values generated by the SHAP algorithm are quite difficult to understand unless we make use of intuitive visualizations. The choice of visualization depends on the choice of global interpretability or local interpretability, which we will cover in this section.

Global interpretability with feature importance bar plots

Analyzing the most influential features present in a dataset always helps us to understand the functioning of an algorithm with respect to the underlying data. SHAP provides an effective way in which to find feature importance using Shapley values. So, the feature importance bar plot displays the important features in descending order of their importance. Additionally, SHAP provides a unique way to show feature interactions using **hierarchical clustering** (`https://www.displayr.com/what-is-hierarchical-clustering/`). These feature clustering methods help us to visualize a group of features that collectively impacts the model's outcome. This is very interesting since one of the core benefits of using Shapley values is to analyze the additive influence of multiple features together. However, there is one drawback of the feature importance plot for global interpretability. Since this method only considers mean absolute Shapley values to estimate feature importance, it doesn't show whether collectively certain features are impacting the model in a negative way or not.

The following diagram illustrates the visualizations for a feature importance plot and a feature clustering plot using SHAP:

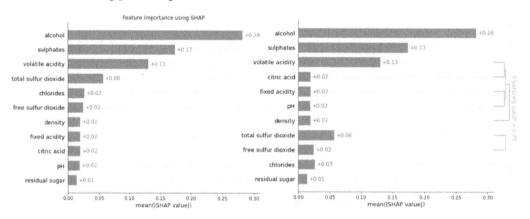

Figure 6.6 – A feature importance plot for global interpretability (left-hand side) and a feature clustering plot for global interpretability (right-hand side)

Next, let's explore SHAP Cohort plots.

Global interpretability with the Cohort plot

Sometimes, analyzing subgroups of data is an important part of data analysis. SHAP provides a very interesting way of grouping the data into certain defined cohorts to analyze feature importance. I found this to be a unique option in SHAP, which can be really helpful! This is an extension of the existing feature importance visualization, and it highlights feature importance for each of the cohorts for a better comparison.

Figure 6.7 shows us the cohort plot to compare two cohorts that are defined from the data:

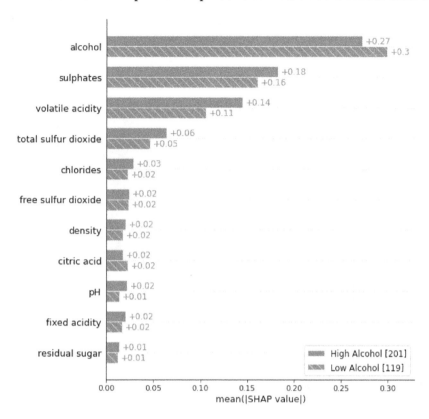

Figure 6.7 – A cohort plot visualization to compare feature importance between two cohorts

Next, we will explore SHAP heatmap plots.

Global interpretability with heatmap plots

To understand the overall impact of all features on the model at a more granular level, heatmap visualizations are extremely useful. The SHAP heatmap visualization shows how every feature value can positively or negatively impact the outcome. Additionally, the plot also includes a line plot to show how the model prediction varies with the positive or negative impact of feature values. However, for non-technical users, this visualization can be really challenging to interpret. This is one of the drawbacks of this visualization method.

Figure 6.8 illustrates a *SHAP heatmap visualization*:

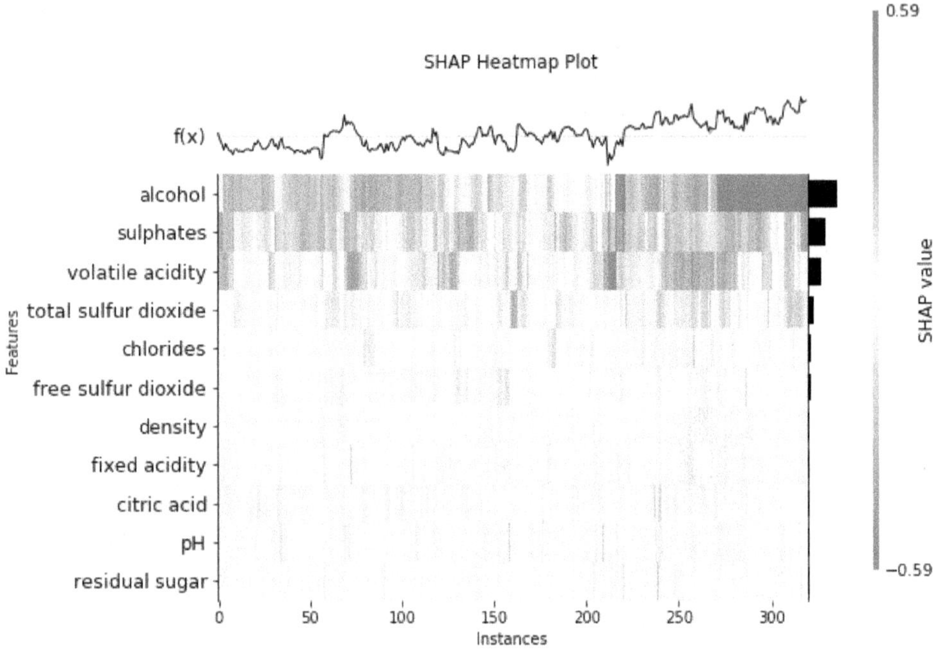

Figure 6.8 – A SHAP heatmap plot

Another popular choice of visualization for global interpretability using SHAP is summary plots. Let's discuss summary plots in the next section.

Global interpretability with summary plots

A summary plot is another visualization method in SHAP for providing global explainability of black-box models. It is a good replacement for the feature importance plot, which not only includes the important features but also the range of effects of these features present in the dataset. The color bar indicates the impact of the features. The features that influence the model's outcome in a positive way are highlighted in a particular color, whereas the features that impact the model's outcome negatively are represented in another contrasting color. The horizontal violin plot for each feature shows the distribution of the Shapley values of the features for each data instance.

The following screenshot illustrates a *SHAP summary plot*:

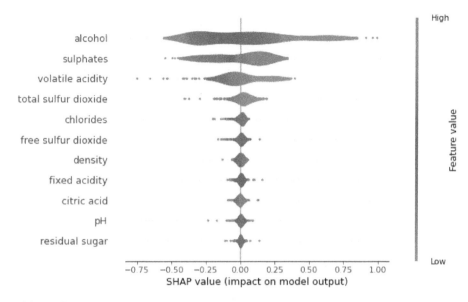

Figure 6.9 – A SHAP violin summary plot

In the next section, we will discuss SHAP dependence plots.

Global interpretability with dependence plots

In certain scenarios, it is important to analyze interactions between the features and how this interaction influences the model outcome. So, SHAP feature dependence plots show the variation of the model outcome by specific features. These plots are similar to the *partial dependence plots*, which were covered in *Chapter 2, Model Explainability Methods*. This plot can help to pick up interesting interaction patterns or trends between the feature values. The features used for selecting the color map are automatically picked up by the algorithm, based on the interaction with a specific selected feature.

Figure 6.10 illustrates a SHAP dependence plot:

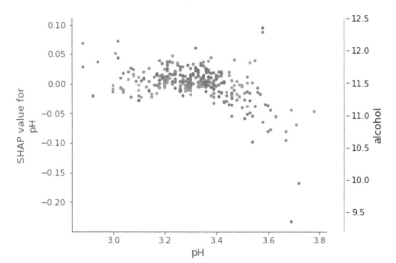

Figure 6.10 – A SHAP dependence plot for the pH feature

In this example, the selected feature is *pH*, and the feature used for selecting the colormap is *alcohol*. So, the plot tells us that with an increase in *pH*, the *alcohol* value also increases. This will be covered in greater detail in the next section.

In the next section, let's explore the SHAP visualization methods used for local explainability.

Local interpretability with bar plots

So far, we have covered various visualization techniques offered by SHAP for providing a global overview of the model. However, similar to LIME, SHAP is also model-agnostic that is designed to provide local explainability. SHAP provides certain visualization methods that can be applied to inference data for local explainability. Local feature importance using SHAP bar plots is one such local explainability method. This plot can help us analyze the positive and negative impact of the features that are present in the data. The features that impact the model's outcome positively are highlighted in one color (pinkish-red by default), and the features having a negative impact on the model outcome are represented using another color (blue by default). Also, as we have discussed before, if the Shapley value is zero for any feature, this indicates that the feature does not influence the model outcome at all. Additionally, the bar plot centers at zero to show the contribution of the features present in the data.

The following diagram shows a *SHAP feature importance bar plot* for local interpretability:

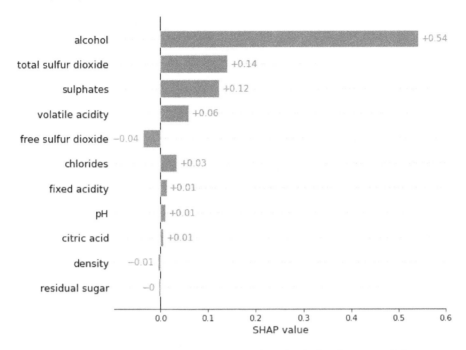

Figure 6.11 – A SHAP feature importance bar plot for local interpretability

Next, let's cover another SHAP visualization that is used for local interpretability.

Local interpretability with waterfall plots

Bar charts are not the only visualization provided by SHAP for local interpretability. The same information can be displayed using a waterfall plot, which might look more attractive. Perhaps, the only difference is that waterfall plots are not centered at zero, whereas bar plots are centered at zero. Otherwise, we get the same feature importance based on Shapley values and the positive or the negative impact of the specific features on the model outcome.

Figure 6.12 illustrates a *SHAP waterfall plot* for local interpretability:

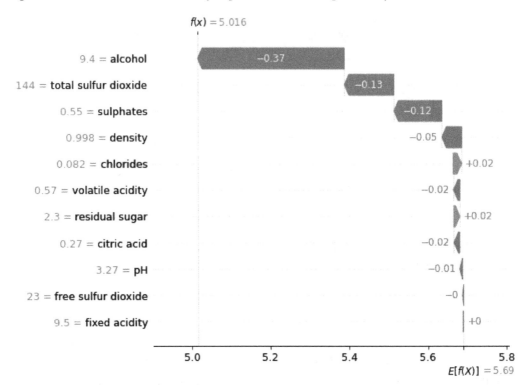

Figure 6.12 – A SHAP waterfall plot for local interpretability

Next, we will discuss force plot visualization in SHAP.

Local interpretability with force plot

We can also use force plots instead of waterfall or bar plots to explain local inference data. With force plots, we can see the model prediction, which is denoted by *f(x)*, as shown in *Figure 6.13*. The *base value* in the following diagram represents the average predicted outcome of the model. The base value is actually used when the features present in the local data instance are not considered. So, using the force plot, we can also see how far the predicted outcome is from the base value. Additionally, we can see the feature impacts as the visual highlights certain features that try to increase the model prediction (which is represented in pink in *Figure 6.13*) along with other important features that have a negative influence on the model as it tries to lower the prediction value (which is represented in green in *Figure 6.13*).

So, *Figure 6.13* illustrates a sample force plot visualization in SHAP:

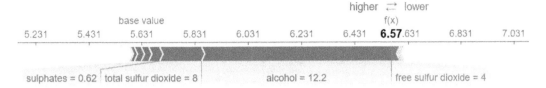

Figure 6.13 – A SHAP force plot for local interpretability

Although force plots might visually look very interesting, we would recommend using bar plots or waterfall plots if the dataset contains many features affecting the model outcome in a positive or a negative way.

Local interpretability with decision plots

The easiest way to explain something is by comparing it with a reference value. So far, in bar plots, waterfall plots, and even force plots, we do not see any reference values for the underlying features used. However, in order to find out whether the feature values are positively or negatively influencing the model outcome, the algorithm is actually trying to compare the feature values of the inference data with the mean of the feature values of the trained model. So, this is the reference value that is not displayed in the three local explainability visualization plots that we covered. But SHAP decision plots help us to compare the feature values of the local data instance with the mean feature values of the training dataset. Additionally, decision plots show the deviation of the feature values, the model prediction, and the direction of deviation of features from the reference values. If the direction of deviation is toward the right, this indicates that the feature is positively influencing the model outcome; if the direction of deviation is toward the left, this represents the negative influence of the feature on the model outcome. Different colors are used to highlight positive or negative influences. If there is no deviation, then the features are actually not influencing the model outcome.

The following diagram illustrates the use of decision plots to compare two different data instances for providing local explainability:

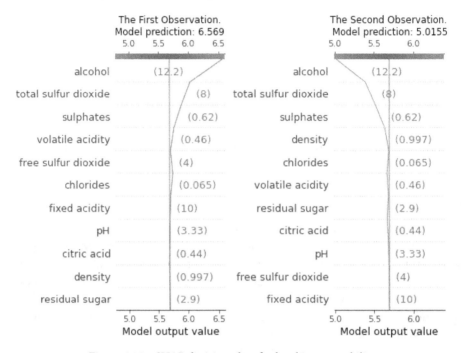

Figure 6.14 – SHAP decision plots for local interpretability

So far, you have seen the variety of visualization methods provided in SHAP for the global and local explainability of ML models. Now, let's discuss the various types of explainers in SHAP.

Explainers in SHAP

In the previous section, we looked at how the data visualization techniques that are available in SHAP can be used to provide explainability. But the choice of the visualization method might also depend on the choice of the explainer algorithm. As we discussed earlier, SHAP provides both model-specific and model-agnostic explainability. But the framework has multiple explainer algorithms that can be applied with different models and with different types of datasets. In this section, we will cover the various explainer algorithms provided in SHAP.

TreeExplainer

TreeExplainer is a fast implementation of the **Tree SHAP algorithm** (`https://arxiv.org/pdf/1802.03888.pdf`) for computing Shapley values for trees and tree-based ensemble learning algorithms. The algorithm makes many diverse possible assumptions about the feature dependence of the features present in the dataset. Only tree-based algorithms are supported such as *Random Forest*, *XGBoost*, *LightGBM*, and *CatBoost*. The algorithm relies on fast C++ implementations in either the local compiled C extension or inside an external model package, but it is faster than conventional Shapley value-based explainers. Generally, it is used for tree-based models trained on structured data for both classification and regression problems.

DeepExplainer

Similar to LIME, SHAP can also be applied to deep learning models trained on unstructured data, such as images and texts. SHAP uses DeepExplainer, which is based on the **Deep SHAP algorithm** for explaining deep learning models. The DeepExplainer algorithm is designed for deep learning models to approximate SHAP values. The algorithm is a modified version of the **DeepLIFT algorithm** (`https://arxiv.org/abs/1704.02685`). The developers of the framework have mentioned that the implementation of the Deep SHAP algorithm differs slightly from the original DeepLIFT algorithm. It uses a distribution of background samples rather than a single reference value. Additionally, the Deep SHAP algorithm also uses Shapley equations to linearize computations such as products, division, max, softmax, and more. The framework mostly supports deep learning frameworks such as TensorFlow, Keras, and PyTorch.

GradientExplainer

DeepExplainer is not the only explainer in SHAP that can be used with deep learning models. GradientExplainer can also work with deep learning models. The algorithm explains models using the concept of **expected gradients**. The expected gradient is an extension of **Integrated Gradients** (`https://arxiv.org/abs/1703.01365`), SHAP, and **SmoothGrad** (`https://arxiv.org/abs/1706.03825`), which combines the ideas of these algorithms into a single expected value equation. Consequently, similar to DeepExplainer, the entire dataset can be used as the background distribution sample instead of a single reference sample. This allows the model to be approximated with a linear function between individual samples of the data and the current input data instance that is to be explained. Since the input features are assumed to be independent, the expected gradients will calculate the approximate SHAP values.

For model explainability, the feature values with higher SHAP values are highlighted, as these features have a positive contribution toward the model's outcome. For unstructured data such as images, pixel positions that have the maximum contribution toward the model prediction are highlighted. Usually, GradientExplainer is slower than DeepExplainer as it makes different approximation assumptions.

The following diagram shows a sample GradientExplainer visualization for the local explainability of a classification model trained on images:

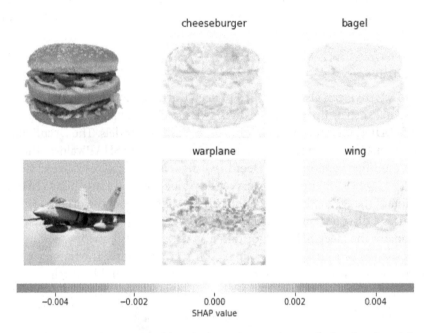

Figure 6.15 – Visualizations used by SHAP GradientExplainers for local explainability

Next, let's discuss SHAP KernelExplainers.

KernelExplainer

KernelExplainers in SHAP use the **Kernel SHAP** method to provide model-agnostic explainability. To estimate the SHAP values for any model, the Kernel SHAP algorithm utilizes a specifically weighted local linear regression approach to compute feature importance. The approach is similar to the LIME algorithm that we have discussed in *Chapter 4, LIME for Model Interpretability*. The major difference between Kernel SHAP and LIME is the approach that is adopted to assign weights to the instances in a regression model.

In LIME, the weights are assigned based on how close the local data instances are to the original instance. Whereas in Kernel SHAP, the weights are assigned based on the estimated Shapley values of the coalition of features used. In simple words, LIME assigns weights based on isolated features, whereas SHAP considers the combined effect of the features for assigning the weights. KernelExplainer is slower than model-specific algorithms as it does not make any assumptions about the model type.

LinearExplainer

SHAP LinearExplainer is designed for computing SHAP values for linear models to analyze inter-feature correlations. LinearExplainer also supports the estimation of the feature covariance matrix for coalition feature importance. However, finding feature correlations for a high-dimensional dataset can be computationally expensive. But LinearExplainers are fast and efficient as they use sampling to estimate a transformation. This is then used for explaining any outcome of linear models.

Therefore, we have discussed the theoretical aspect of various explainers in SHAP. For more information about these explainers, I do recommend checking out `https://shap-lrjball.readthedocs.io/en/docs_update/api.html`. In the next chapter, we will cover the practical implementation of the SHAP explainers using the code tutorials on GitHub in which we will implement the SHAP explainers for explaining models trained on different types of datasets. In the next section, we will cover a practical tutorial on how to use SHAP for explaining regression models to give you a glimpse of how to apply SHAP for model explainability.

Using SHAP to explain regression models

In the previous section, we learned about different visualizations and explainers in SHAP for explaining ML models. Now, I will give you practical exposure to using SHAP for providing model explainability. The framework is available as an open source project on GitHub: `https://github.com/slundberg/shap`. You can get the API documentation at `https://shap-lrjball.readthedocs.io/en/docs_update/index.html`. The complete tutorial is provided in the GitHub repository at `https://github.com/PacktPublishing/Applied-Machine-Learning-Explainability-Techniques/blob/main/Chapter06/Intro_to_SHAP.ipynb`. I strongly recommend that you read this section and execute the code side by side.

Setting up SHAP

Installing SHAP in Python can be done easily using the pip installer by using the following command in your console:

```
pip install shap
```

Since the tutorial requires you to have other Python frameworks installed, you can also try the following command to install all the necessary modules for the tutorial from the Jupyter notebook itself if it is not installed already:

```
!pip install --upgrade pandas numpy matplotlib seaborn scikit-learn shap
```

Now, let's import SHAP and check its version:

```
import shap
print(f"Shap version used: {shap.__version__}")
```

The version that I have used for this tutorial is *0.40.0*.

Please note that with different versions, there can be different changes to the API or different errors that you might face. So, I will recommend you to look at the latest documentation of the framework if you encounter any such issues. I have also added a **SHAP Errata** (https://github.com/PacktPublishing/Applied-Machine-Learning-Explainability-Techniques/tree/main/Chapter06/SHAP_ERRATA) in the repository for providing solutions to existing known issues with the SHAP framework.

Inspecting the dataset

For this tutorial, we will use the *Red Wine Quality Dataset* from *Kaggle*: https://www.kaggle.com/uciml/red-wine-quality-cortez-et-al-2009. The dataset has already been added to the code repository so that it is easy for you to access the data. This particular dataset contains information about the red variant of the Portuguese *Vinho Verde* wine, and it is derived from the original UCI source at https://archive.ics.uci.edu/ml/datasets/wine+quality.

> **Wine Quality Data Set**
>
> The credit for this dataset goes to *P. Cortez, A. Cerdeira, F. Almeida, T. Matos and J. Reis. Modeling wine preferences by data mining from physicochemical properties.*

We will use this dataset to solve a regression problem. We will load the data as a pandas DataFrame and perform an initial inspection:

```
data = pd.read_csv('dataset/winequality-red.csv')
data.head()
```

Figure 6.16 illustrates a snapshot of the data:

fixed acidity	volatile acidity	citric acid	residual sugar	chlorides	free sulfur dioxide	total sulfur dioxide	density	pH	sulphates	alcohol	quality
7.4	0.70	0.00	1.9	0.076	11.0	34.0	0.9978	3.51	0.56	9.4	5
7.8	0.88	0.00	2.6	0.098	25.0	67.0	0.9968	3.20	0.68	9.8	5
7.8	0.76	0.04	2.3	0.092	15.0	54.0	0.9970	3.26	0.65	9.8	5
11.2	0.28	0.56	1.9	0.075	17.0	60.0	0.9980	3.16	0.58	9.8	6
7.4	0.70	0.00	1.9	0.076	11.0	34.0	0.9978	3.51	0.56	9.4	5

Figure 6.16 – Snapshot of the pandas DataFrame of the Wine Quality dataset

We can check some quick information about the dataset using the following command:

```
data.info()
```

This gives the following output:

```
RangeIndex: 1599 entries, 0 to 1598
Data columns (total 12 columns):
 #   Column                Non-Null Count  Dtype
---  ------                --------------  -----
 0   fixed acidity         1599 non-null   float64
 1   volatile acidity      1599 non-null   float64
 2   citric acid           1599 non-null   float64
 3   residual sugar        1599 non-null   float64
 4   chlorides             1599 non-null   float64
 5   free sulfur dioxide   1599 non-null   float64
 6   total sulfur dioxide  1599 non-null   float64
 7   density               1599 non-null   float64
 8   pH                    1599 non-null   float64
 9   sulphates             1599 non-null   float64
 10  alcohol               1599 non-null   float64
 11  quality               1599 non-null   int64
dtypes: float64(11), int64(1)
memory usage: 150.0 KB
```

As you can see, our dataset consists of *11 numerical features* and *1,599* records of data. The target outcome that the regression model will be learning is the *quality* of the wine, which is an *integer feature*. Although we are using this dataset for solving a regression problem, the same problem can be viewed as a classification problem and the same underlying data can be used.

The purpose of the tutorial is not for building an extremely efficient model, but rather to consider any model and use SHAP to explain the workings of the model. So, we will skip the EDA, data normalization, outlier detection, and even feature engineering steps, which are, otherwise, crucial steps for building a robust ML model. But missing values in the dataset can create problems for the SHAP algorithm. Therefore, I would suggest doing a quick check of missing values at the very least:

```
sns.displot(
    data=data.isna().melt(value _ name="missing"),
    y="variable",
    hue="missing",
    multiple="fill",
    aspect=1.5
)
plt.show()
```

The preceding lines of code will result in the following plot as its output:

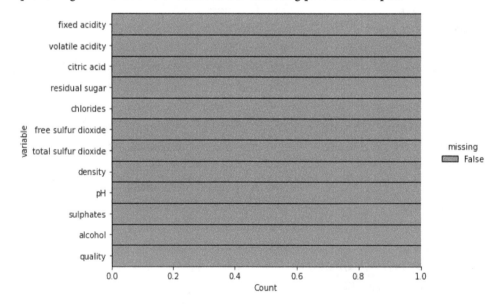

Figure 6.17 – The missing plot visualization for the dataset

Fortunately, the dataset doesn't have any missing values; otherwise, we might have to handle this before proceeding further. But we are good to proceed with the modeling step as there are no significant issues with the data.

Training the model

Since I do not have any pretrained model for this dataset, I thought of building a simple random forest model. We can divide the model into a training and a testing set using the 80:20 split ratio:

```
features = data.drop(columns=['quality'])
labels = data['quality']
# Dividing the data into training-test set with 80:20 split ratio
x_train,x_test,y_train,y_test = train_test_split(
    features,labels,test_size=0.2, random_state=123)
```

To use the random forest algorithm, we would need to import this algorithm from the scikit-learn module and then fit the regression model on the training data:

```
from sklearn.ensemble import RandomForestRegressor
model = RandomForestRegressor(n_estimators=2000,
                             max_depth=30,
                             random_state=123)
model.fit(x_train, y_train)
```

> **Important Note**
>
> Considering the objective of this notebook, we are not doing an extensive hyperparameter tuning process. But I would strongly recommend you to perform all necessary best practices like *EDA, feature engineering, hyperparameter tuning, cross-validation*, and others for your use cases.

Once the trained model is ready, we will do a quick evaluation of the model using the metric of **coefficient of determination (R2 coefficient)**:

```
model.score(x_test, y_test)
```

The model score obtained is just around 0.5, which indicates that the model is not very efficient. So, model explainability is even more important for such models. Now, let's use SHAP to explain the model.

Application of SHAP

Applying SHAP is very easy and can be done in a few lines of code. First, we will use the Shapley value-based explainer on the test dataset:

```
explainer = shap.Explainer(model)
shap_values = explainer(x_test)
```

Then, we can use the SHAP values for the various visualization techniques discussed earlier. The choice of the visualizations depends on whether we want to go for global explainability or local explainability. For example, for the summary plot shown in *Figure 6.9*, we can use the following code:

```
plt.title('Feature Importance using SHAP')
shap.plots.bar(shap_values, show=True, max_display=12)
```

To provide local explainability, if we want to use the decision plot shown in *Figure 6.14*, we can try the following code:

```
expected_value = explainer.expected_value
shap_values = explainer.shap_values(x_test)[0]
shap.decision_plot(expected_value, shap_values, x_test)
```

To use a different explainer algorithm, we just need to select the appropriate explainer. For Tree Explainers, we can try the following code:

```
explainer = shap.TreeExplainer(model)
shap_values = explainer.shap_values(x_test)
```

The usage of this framework to explain the regression model while considering various aspects has already been covered in the notebook tutorial: https://github.com/PacktPublishing/Applied-Machine-Learning-Explainability-Techniques/blob/main/Chapter06/Intro_to_SHAP.ipynb.

In the next chapter, we will cover more interesting use cases. Next, let's discuss some advantages and disadvantages of this framework.

Advantages and limitations of SHAP

In the previous section, we discussed the practical application of SHAP for explaining a regression model with just a few lines of code. However, since SHAP is not the only explainability framework, we should be aware of the specific advantages and disadvantages of SHAP, too.

Advantages

The following is a list of some of the advantages of SHAP:

- **Local explainability**: Since SHAP provides local explainability to inference data, it enables users to analyze key factors that are positively or negatively affecting the model's decision-making process. As SHAP provides local explainability, it is useful for production-level ML systems, too.

- **Global explainability**: Global explainability provided in SHAP helps to extract key information about the model and the training data, especially from the collective feature importance plots. I think SHAP is better than LIME for getting a global perspective on the model. SP-LIME in LIME is good for getting an example-driven global perspective of the model, but I think SHAP provides a generalized global understanding of trained models.

- **Model-agnostic and model-specific**: SHAP can be model-agnostic and model-specific. So, it can work with black-box models and also work with complex deep learning models to provide explainability.

- **Theoretical robustness**: The concept of using Shapley values for model explainability, which is based on the principles of coalition game theory, captures feature interaction very well. Also, the properties of SHAP regarding *efficiency*, *symmetry*, *dummy*, and *additivity* are formulated on a robust theoretical foundation. Unlike SHAP, LIME is not based on a solid theory as it assumes ML models will behave linearly for some local data points. But there is not much theoretical evidence that proves why this assumption is true for all cases. That is why I would say SHAP is based on ideas that are theoretically more robust than LIME.

These advantages make SHAP one of the most popular choices of the XAI framework. Unfortunately, applying SHAP can be really challenging for high-dimensional datasets as it does not provide actionable explanations. Let's look at some of the limitations of SHAP.

Limitations

Here is a list of some of the limitations of SHAP:

- **SHAP is not the preferred choice for high-dimensional data**: Computing Shapley values on high-dimensional data can be computationally more challenging, as the time complexity of the algorithm is $O(2^n)$, where n is the total number of features in the dataset.

- **Shapley values are ineffective for selective explanation**: Shapley values try to consider all the features for providing explainability. The explanations can be incorrect for sparse explanations, in which only selected features are considered. But usually, human-friendly explanations consider selective features. So, I would say that LIME is better than SHAP when you seek a selective explanation. However, more recent versions of the SHAP framework do include the same ideas as LIME and can be almost equally effective for sparse explanations.

- **SHAP cannot be used for prescriptive insights**: SHAP computes the Shapley values for each feature and does not build a prediction model like LIME. So, it cannot be used for analyzing any *what-if scenario* or for providing any *counterfactual example* for suggesting actionable insights.

- **KernelSHAP can be slow**: Although KernelSHAP is model-agnostic, it can be very slow and, thus, might not be suitable for production-level ML systems for models trained on high-dimensional data.

- **Not extremely human-friendly**: Apart from analyzing feature importance through feature interactions, SHAP visualizations can be complicated to interpret for any non-technical user. Often, non-technical users prefer simple selective actionable insights, recommendations, or justifications from ML models. Unfortunately, SHAP requires another layer of abstraction for human-friendly explanations when used in production systems.

As we can see from the points discussed in this section, SHAP might not be the most ideal framework for explainability, and there is a lot of space for improvement to make it more human-friendly. However, it is indeed an important and very useful framework for explaining black-box algorithms, especially for technical users. This brings us to the end of the chapter. Let's summarize what we have learned in the chapter.

Summary

In this chapter, we focused on understanding the importance of the SHAP framework for model explainability. By now, you have a good understanding of Shapley values and SHAP. We have covered how to use SHAP for model explainability through a variety of visualization and explainer methods. Also, we have covered a code walk-through for using SHAP to explain regression models. Finally, we discussed some advantages and limitations of the framework.

In the next chapter, we will cover more interesting practical use cases for applying SHAP on different types of datasets.

References

For additional information, please refer to the following resources:

- *Shapley, Lloyd S. "A Value for n-Person Games." Contributions to the Theory of Games 2.28 (1953)*: `https://doi.org/10.1515/9781400881970-018`

- The Red Wine Quality dataset from Kaggle: `https://www.kaggle.com/uciml/red-wine-quality-cortez-et-al-2009`

- The SHAP GitHub project: `https://github.com/slundberg/shap`

- The official SHAP documentation: `https://shap.readthedocs.io/en/latest/index.html`

7

Practical Exposure to Using SHAP in ML

In the previous chapter, we discussed **SHapley Additive exPlanation (SHAP)**, which is one of the most popular model explainability frameworks. We also covered a practical example of using SHAP for explaining regression models. However, SHAP can explain other types of models trained on different types of datasets. In the previous chapter, you did receive a brief conceptual understanding of the different types of **explainers** available in SHAP for explaining models trained on different types of datasets. But in this chapter, you will get the practical exposure needed to apply the various types of explainers available in SHAP.

More specifically, you learn how to apply **TreeExplainers** for explaining tree ensemble models trained on structured tabular data. You will also learn how to apply **DeepExplainer** and **GradientExplainer** SHAP with deep learning models trained on image data. As you learned in the previous chapter, the **KernelExplainer** in SHAP is model-agnostic, and you will get practical exposure to KernelExplainers in this chapter. We will also cover the practical aspect of using **LinearExplainers** on linear models. Finally, you will get to explore how SHAP can be used to explain the complicated state-of-the-art **Transformer** models trained on text data.

In this chapter, we will cover the following important topics:

- Applying TreeExplainers to tree ensemble models
- Explaining deep learning models using DeepExplainer and GradientExplainer
- Model-agnostic explainability using KernelExplainer
- Exploring LinearExplainer in SHAP
- Explaining transformers using SHAP

Let's get started!

Technical requirements

This code tutorial along with the necessary resources can be downloaded or cloned from the GitHub repository for this chapter: `https://github.com/PacktPublishing/Applied-Machine-Learning-Explainability-Techniques/tree/main/Chapter07`. Python and Jupyter notebooks are used to implement the practical application of the theoretical concepts covered in this chapter. However, I will recommend that you run the notebooks only after you go through this chapter for a better understanding.

Applying TreeExplainers to tree ensemble models

As discussed in the previous chapter, the Tree SHAP implementation can work with tree ensemble models such as **Random Forests**, **XGBoost**, and **LightGBM** algorithms. Now, decision trees are inherently interpretable. But tree-based ensemble learning models, either implementing boosting or bagging, are not inherently interpretable and can be quite complex to interpret. So, SHAP is one of the popular choices of algorithms used to explain such complex models. The Kernel SHAP implementation of SHAP is model-agnostic and can explain any model. However, the algorithm can be really slow with larger datasets with many features. That is why the **Tree SHAP** (`https://arxiv.org/abs/1802.03888`) implementation of the algorithm is a high-speed exact algorithm for tree ensemble models. TreeExplainer is the fast C++ implementation of the Tree SHAP algorithm, which supports algorithms such as XGBoost, CatBoost, LightGBM, and other tree ensemble models from scikit-learn. In this section, I will cover how to apply TreeExplainer in practice.

Installing the required Python modules

The complete tutorial is provided in the GitHub repository at `https://github.com/PacktPublishing/Applied-Machine-Learning-Explainability-Techniques/blob/main/Chapter07/TreeExplainers.ipynb` and I strongly recommend that you read this section and execute the code side by side. If you have followed the previous tutorials provided in the other chapters, most of the required Python packages should be installed by now. Otherwise, you can install the necessary packages using the `pip` installer:

```
!pip install --upgrade numpy pandas matplotlib seaborn sklearn
lightgbm shap
```

You can import these packages to verify their successful installations:

```
import numpy as np
import pandas as pd
import seaborn as sns
import matplotlib.pyplot as plt
import sklearn
import lightgbm as lgb
import shap
```

For certain JavaScript-based SHAP visualizations in Jupyter notebooks, make sure to initialize the SHAP JavaScript module:

```
shap.initjs()
```

Next, let's discuss the dataset that we are going to use for this example.

Discussion about the dataset

For this example, we will use the German Credit Risk dataset from Kaggle (`https://www.kaggle.com/uciml/german-credit`). This dataset is used to build a classification model for classifying good and bad credit risk. The Kaggle dataset is actually a simplified version of the original data available in UCI (`https://archive.ics.uci.edu/ml/datasets/statlog+(german+credit+data)`)

> **Statlog (German Credit Data) Data Set**
>
> The credit for the dataset goes to *Professor Dr. Hans Hofmann, Institut für Statistik und Ökonometrie Universität Hamburg.*

Please refer to the notebook for more information on the dataset. The dataset is already provided in the GitHub repository for this chapter: `https://github.com/PacktPublishing/Applied-Machine-Learning-Explainability-Techniques/tree/main/Chapter07/datasets`. We can use the pandas Python module to load and display the dataset as a DataFrame:

```
data  = pd.read_csv('datasets/german_credit_data.csv', index_
col=0)
data.head()
```

The following diagram illustrates the pandas DataFrame for this data:

Age	Sex	Job	Housing	Saving accounts	Checking account	Credit amount	Duration	Purpose	Risk
67	male	2	own	NaN	little	1169	6	radio/TV	good
22	female	2	own	little	moderate	5951	48	radio/TV	bad
49	male	1	own	little	NaN	2096	12	education	good
45	male	2	free	little	little	7882	42	furniture/equipment	good
53	male	2	free	little	little	4870	24	car	bad

Figure 7.1 – pandas DataFrame snapshot of the German Credit Risk dataset

I do recommend that you perform a thorough **Exploratory Data Analysis (EDA)**. You can also use pandas profiling (`https://github.com/ydataai/pandas-profiling`) as shown in *Chapter 2, Model Explainability Methods*, for automated EDA. Since we have covered this already, I will skip the EDA part for this example.

However, the dataset does have some missing values, which needs to be handled before building a model. We can check that using the following lines of code:

```
sns.displot(
    data=data.isna().melt(value_name="missing"),
    y="variable",
    hue="missing",
    multiple="fill",
    aspect=1.5,
    palette='seismic'
)
plt.show()
```

The following visualization is generated as output:

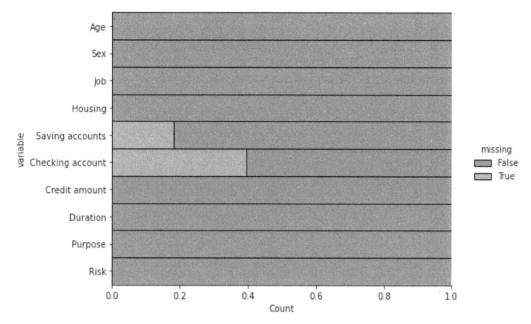

Figure 7.2 – Missing value visualization for the German Credit Risk dataset

The dataset has around 18% missing values for the `Saving accounts` feature and 40% missing values for the `Checking account` feature. Since the percentage of missing data is high, and the features can be important, we cannot completely ignore or drop these features. Please remember that the focus of this tutorial is on model explainability using TreeExplainers. So, we will not spend too much time doing data imputation as we are not concerned with building an efficient model for this example. As both the features are categorical features, we will simply create an `Unknown` category for the missing values. This can be done by means of the following line of code:

```
data.fillna('Unknown', inplace=True)
```

We will need to perform **Label Encoding** for the categorical features as we need to convert the string-like feature values to an integer format:

```
from sklearn.preprocessing import LabelEncoder

le = LabelEncoder()
for feat in ['Sex', 'Housing', 'Saving accounts', 'Checking
account', 'Purpose', 'Risk']:
```

```
le.fit(data[feat])
data[feat] = le.transform(data[feat])
```

Now, for this example, we will use the **LightGBM algorithm** (`https://lightgbm.readthedocs.io/en/latest/`), which can work directly on categorical variables, and hence we do not need to perform **one-hot encoding**. But for other algorithms, we might need to perform one-hot encoding. Moreover, we will not perform other complex data pre-processing or feature engineering steps. I do recommend that you perform robust *feature engineering*, *outlier detection*, and *data normalization* for building efficient ML models. However, for this example, even if the model is not very accurate, we can use SHAP to generate explanations. Let's proceed to the model training part.

Training the model

Before training the model, we will need to create the training and test sets:

```
from sklearn.model_selection import train_test_split

features = data.drop(columns=['Risk'])
labels = data['Risk']
# Dividing the data into training-test set with 80:20 split
ratio
x_train,x_test,y_train,y_test = \
train_test_split(features,labels,test_size=0.2,
                 random_state=123)
```

Since we will be using the LightGBM algorithm, we will need to create LightGBM dataset objects, which are used during the training process:

```
data_train = lgb.Dataset(x_train, label=y_train,
                         categorical_feature=cat_features)
data_test = lgb.Dataset(x_test, label=y_test,
                        categorical_feature=cat_features)
```

We also need to define the model parameters as a dictionary:

```
params = {
    'boosting_type': 'gbdt',
    'objective': 'binary',
    'metric': 'auc',
```

```
    'num_leaves': 20,
    'learning_rate': 0.05,
    'feature_fraction': 0.9,
    'bagging_fraction': 0.8,
    'bagging_freq': 5,
    'verbose': -1,
    'lambda_l1': 1,
    'lambda_l2': 1,
    'seed': 123
}
```

Finally, we can train the model using the parameters and dataset object created:

```
model = lgb.train(params,
                  data_train,
                  num_boost_round=100,
                  verbose_eval=100,
                  valid_sets=[data_test, data_train])
```

We are again skipping the hyperparameter tuning process to obtain a more efficient model, but I would definitely recommend spending some time on hyperparameter tuning to get a model with higher accuracy. Now, let's proceed to the model explainability part using SHAP.

Application of TreeExplainer in SHAP

Applying TreeExplainer in SHAP is very easy as the framework is well modularized:

```
explainer = shap.TreeExplainer(model)
shap_values = explainer.shap_values(features)
```

Once we have approximated the SHAP values, we can then apply the visualization methods provided in SHAP to obtain the model's explainability. I would recommend that you refer to the *Visualizations in SHAP* section in *Chapter 6, Model Interpretability Using SHAP*, to refresh your memory regarding the various visualization methods that we can use with SHAP for model explainability. We will start with global explainability with summary plots.

Figure 7.3 illustrates the SHAP summary plot using the SHAP values generated by TreeExplainer on this dataset:

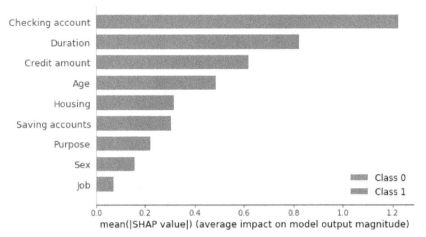

Figure 7.3 – SHAP summary plot on SHAP values generated by TreeExplainer

As we can see from the preceding figure, the summary plot highlights the important features based on SHAP values, ordered from most important to least important. The model considered Checking account and Duration as one of the most influential features, compared to the Sex or Job features.

For local explainability, we can apply the **force plot** and **decision plot** visualization methods:

```
# Local explainability with force plots
shap.force_plot(explainer.expected_value[1], shap_values[1]
[0,:], features.iloc[0,:])
# Local explainability with force plots
shap.decision_plot(explainer.expected_value[1], shap_values[1]
[0,:], features.iloc[0,:])
```

I often find decision plots to be more interpretable than force plots as decision plots show you the deviation from the mean expected value for each feature. The direction of deviation also indicates whether the feature is positively impacting the model's decision or whether it has a negative impact. But some of you might prefer force plots as well, as this indicates the positively or negatively affecting features based on their feature values and how they can impact in terms of achieving a higher prediction value or a lower prediction value.

Figure 7.4 illustrates the force plot and decision plot that we have obtained:

Force plot

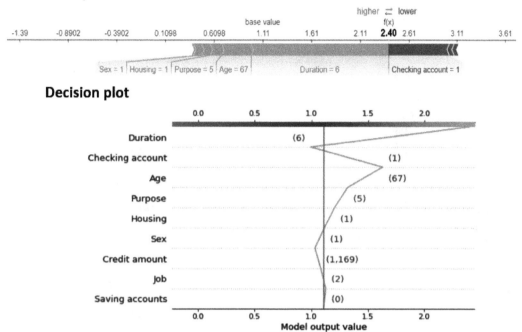

Figure 7.4 – Force and decision plots for local interpretability

In certain cases, understanding the inter-feature dependence becomes important as SHAP doesn't consider features in isolation to obtain the most influential features. Rather SHAP-based feature importance is estimated based on the collective impact of multiple features together. So, for analyzing the feature importance, we can try out the SHAP feature dependence plots:

```
# For feature wise global interpretability
for col in ['Purpose', 'Age']:

    print(f"Feature Dependence plot for: {col}")
    shap.dependence_plot(col, shap_values[1], features,
display_features=features)
```

The following diagram shows the feature dependence plot for the `Purpose` and `Age` features:

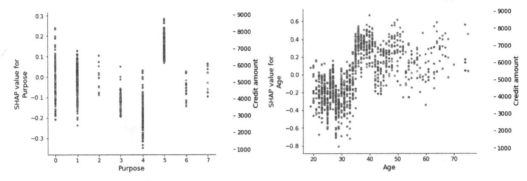

Figure 7.5 – SHAP feature dependence plot for the Purpose and Age features

From the SHAP in *Figure 7.3*, it was slightly surprising for me to find out that the features `Purpose` and `Age` are not as important as `Duration` or `Credit amount`. In such cases, the feature dependence plots automatically calculate the most dependent feature for a selected feature. So, from *Figure 7.5*, we can see that for both `Purpose` and `Age`, `Credit Amount` is the dependent feature, and we can also see how these features vary with the dependent feature. This justifies that collectively, `Credit amount` is more influential than `Purpose` and `Age`.

You can also try out other visualization methods covered in *Chapter 6, Model Interpretability Using SHAP*, and definitely recommend you to play around with the SHAP values generated using the TreeExplainer so that you can come up with your own custom visualization method! In the next section, we are going to apply SHAP explainers to deep learning models trained on image data.

Explaining deep learning models using DeepExplainer and GradientExplainer

In the previous section, we covered the use of TreeExplainer in SHAP, which is a model-specific explainability method only applicable to tree ensemble models. We will now discuss GradientExplainer and DeepExplainer, two other model-specific explainers in SHAP that are mostly used with deep learning models.

GradientExplainer

As discussed in *Chapter 2*, *Model Explainability Methods*, one of the most widely adopted ways to explain deep learning models trained on unstructured data such as images is **layer-wise relevance propagation** (**LRP**). LRP is about analyzing the gradient flow for the intermediate layers of the deep neural network. SHAP GradientExplainers also function in a similar way. As discussed in *Chapter 6*, *Model Interpretability Using SHAP*, GradientExplainer combines the idea of *SHAP*, *integrated gradients*, and *SmoothGrad* into a single expected value equation. GradientExplainer finally uses a sensitivity map-based gradient visualization method. The red pixels in the visualization map represent pixels having positive SHAP values, which increases the probability of the output class. The blue pixels represent pixels having negative SHAP values that decrease the likelihood of the output class. Now, let me walk you through the tutorial provided in the code repository: `https://github.com/PacktPublishing/Applied-Machine-Learning-Explainability-Techniques/blob/main/Chapter07/Explaining%20 Deep%20Learning%20models.ipynb`. Please load the necessary modules and follow the detailed steps provided in the notebook tutorial as I will be covering only the important coding steps in this section for helping you to understand the flow of the code tutorial.

Discussion on the dataset used for training the model

For this example, we will be using the SHAP ImageNet dataset, which will be used to generate the background reference required by the GradientExplainer algorithm. We will also pick up the inference image from the same dataset. However, you are always free to pick up any other image dataset or inference image of your choice:

```
X,y = shap.datasets.imagenet50(resolution=224)
inference_image = X[[46]]
```

For this example, we have selected the following image as our inference image:

Figure 7.6 – Inference image from SHAP ImageNet50

As we can see from the inference image, it contains many possible objects, including a man, chair, and computer. All these can be potential model outcomes and the actual outcome depends on the exact region where the model is looking to make the prediction. So, explainability is very important in such cases. Next, let's discuss the model used for this example.

Using a pre-trained CNN model for this example

I have used a pre-trained CNN model, **VGG19**, as our black-box image classification model. The pre-trained model is trained on ImageNet data, and hence the training and the inference data will be consistent. The model can be loaded using the `tensorflow` Python module:

```
from tensorflow.keras.applications.vgg19 import VGG19
model = VGG19(weights='imagenet')
```

Next, let's apply GradientExplainer to SHAP.

Application of GradientExplainer in SHAP

GradientExplainer helps to map the gradient flow of intermediate layers of a deep learning model such as **Convolution Neural Network (CNN)** to explain the workings of the model. So, we will try to explore the 10th layer of the model and visualize the gradients based on SHAP values. The choice of the 10th layer is completely arbitrary; you can choose other layers as well:

```
layer_num = 10


# explain how the input to the 10th layer of the model explains
the top two classes
def map2layer(x, layer):
    '''

    Source : https://github.com/slundberg/shap
    '''

    feed_dict = dict(zip([model.layers[0].input],
                        [preprocess_input(x.copy())]))
    return K.get_session().run(model.layers[layer].input, feed_
dict)


model_input = (model.layers[layer_num].input,
                model.layers[-1].output)
explainer = shap.GradientExplainer(model_input,
                                    map2layer(X, layer_num),
                                    local_smoothing=0)
shap_values, ind = explainer.shap_values(
    map2layer(inference_image, layer_num),
    ranked_outputs=4)
# plot the explanations
shap.image_plot(shap_values, inference_image, class_names)
```

In this example, we are trying to estimate the GradientExplainer-based SHAP values for the 10th layer of the pre-trained model. Using the SHAP image plot method, we can visualize the sensitivity map for the top 4 probable outcomes of the model, which is illustrated in *Figure 7.7*:

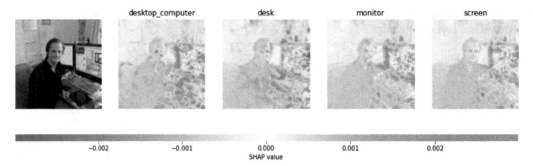

Figure 7.7 – SHAP image plot visualization based on applying GradientExplainer to the inference image

The top four predictions from the model are **desktop_computer**, **desk**, **monitor**, and **screen**. All of these are valid outcomes depending on which region the model is focusing on. Using the SHAP image plots shown in *Figure 7.7*, we can identify the exact regions contributing to the specific model prediction. The pixel regions marked in red are making a positive contribution to the specific model prediction, whereas the blue pixel regions are negatively contributing to the model predictions. You can try visualizing other layers of the model as well and analyze how the model prediction varies throughout the layers!

Exploring DeepExplainers

In the previous section, we covered GradientExplainers in SHAP. However, deep learning models can also be explained using DeepExplainers in SHAP based on the Deep SHAP algorithm. Deep SHAP is a high-speed implementation for estimating SHAP values for deep learning models. It uses a distribution of background samples and Shapley equations to linearize predominant non-linear operations used in deep learning models such as max, products, and softmax.

The tutorial notebook provided in `https://github.com/PacktPublishing/` `Applied-Machine-Learning-Explainability-Techniques/blob/` `main/Chapter07/Explaining%20Deep%20Learning%20models.ipynb` covers an example of a deep learning model trained from scratch on the CIFAR-10 dataset (`https://www.cs.toronto.edu/~kriz/cifar.html`) for multi-class classification. The dataset contains highly compressed images of size 32x32 belonging to 10 different classes. In this section, I will skip the model training process as it is already covered in sufficient detail in the notebook tutorial. Instead, I will discuss the model explainability part using DeepExplainers, which is our primary focus. You can also try out the same tutorial with a pre-trained CNN model instead of training a model from scratch. Now, let's discuss the model explainability part.

Application of DeepExplainer in SHAP

In order to apply DeepExplainer, we need to first form the background samples. The robustness of the explainability actually depends a lot on the selection of the background samples. For this example, we will randomly select 1,000 samples from the training data. You can increase your sample size as well, but please ensure that the background samples have no data drift between the training or the inference data by ensuring that the data collection process is consistent:

```
background = x_train[np.random.choice(len(x_train), 1000,
replace=False)]
```

Once the background samples have been selected, we can create a SHAP explainer object using the DeepExplainer method on the trained model and the background samples and estimate the SHAP values for the inference data:

```
explainer = shap.DeepExplainer(model, background)
shap_values = explainer.shap_values(sample_x_test)
```

After the SHAP values are computed, we can use the SHAP image plot to visualize the pixels influencing the model in both a positive and negative manner using a similar sensitivity plot as used for GradientExplainer:

```
shap.image_plot(shap_values, sample_x_test, labels,
labelpad= 1)
```

The following figure shows the SHAP image plot for some sample inference data:

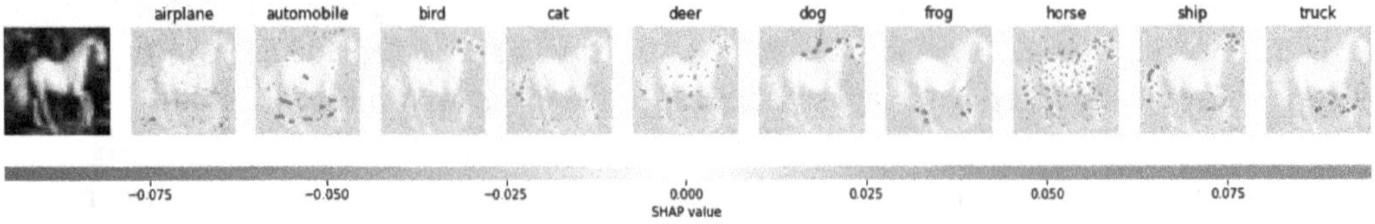

Figure 7.8 – SHAP image plot visualization after applying DeepExplainers in SHAP

As we can see from *Figure 7.8*, even if the model is trained on a very compressed dataset, DeepExplainer was able to calculate SHAP values that can help us identify the regions of the image (highlighted in pinkish-red pixels) that have positively contributed to the model's prediction. The model did correctly predict the outcome as a *horse*, which is the correct classification from the compressed image. However, applying DeepExplainer was quite simple and the method was very fast compared to conventional methods to approximate SHAP values for deep learning models trained on unstructured data such as images. Next, we will learn about KernelExplainer for model agnostic explainability.

Model-agnostic explainability using KernelExplainer

In the previous sections, we have discussed three model-specific explainers available in SHAP – TreeExplainer, GradientExplainer, and DeepExplainer. The KernelExplainer in SHAP actually makes SHAP a model-agnostic explainability approach. However, unlike the previous methods, KernelExplainer based on the Kernel SHAP algorithm is much slower, especially for large and high dimensional datasets. Kernel SHAP tries to combine ideas from Shapley values and **Local Interpretable Model-agnostic Explanations (LIME)** for both global and local interpretability of black-box models. Similar to the approach followed in LIME, the Kernel SHAP algorithm also creates locally linear perturbed samples and computes Shapley values of the same to identify features contributing to or against the model prediction.

KernelExplainer is the practical implementation of the Kernel SHAP algorithm. The complete tutorial demonstrating the application of SHAP KernelExplainer is provided in the following notebook: `https://github.com/PacktPublishing/ Applied-Machine-Learning-Explainability-Techniques/blob/main/ Chapter07/KernelExplainers.ipynb`. I have used the same *German Credit Risk dataset* as used for the *TreeExplainer tutorial*. Please refer to the *Applying TreeExplainers to tree ensemble models* section for a detailed discussion of the dataset and the model if you are starting from this section directly. In this section, we will discuss the application of KernelExplainers for the same problem discussed in the TreeExplainer tutorial.

Application of KernelExplainer in SHAP

The KernelExplainer method in SHAP takes the model and the background data as the input to compute the SHAP values. For larger datasets or high dimensional datasets having many features, it is recommended to use only a subset of the training data as the background samples. Otherwise, Kernel SHAP can be a very slow algorithm and would take a lot of time to generate the SHAP values. Like the previous explainer methods covered, applying KernelExplainer is very simple and can be done using the following lines of code:

```
explainer = shap.KernelExplainer(model.predict, x_train)
shap_values = explainer.shap_values(x_test, nsamples=100)
```

If we log the wall time (using %%time in Jupyter notebooks) for computing SHAP values and compare KernelExplainer with TreeExplainer on the same dataset, we will observe that KernelExplainer takes a significantly longer time to execute (almost 1,000 times longer in our case!). This shows that even though KernelExplainer is model-agnostic, the slowness of the algorithm is a major drawback.

For explaining black-box models, the same visualization methods covered for TreeExplainer are applicable, which can be generated by the following code:

```
shap.summary_plot(shap_values, x_test, plot_type='violin',
show=False)
shap.force_plot(explainer.expected_value, shap_values[1], x_
test.iloc[0,:])
shap.decision_plot(explainer.expected_value, shap_values[1],
x_test.iloc[0,:])
```

Figure 7.9 shows the summary plot, decision plot, and force plots used to explain the black-box model:

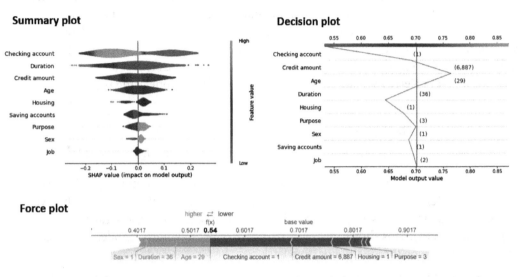

Figure 7.9 – Summary plot, decision plot, and force plots obtained after using SHAP KernelExplainer

The plots shown in *Figure 7.9* can be obtained using the same approach as covered for TreeExplainer. In the next section, we will cover LinearExplainer in SHAP, which is another model-specific explanation method.

Exploring LinearExplainer in SHAP

LinearExplainer in SHAP is particularly developed for linear machine learning models. In the previous section, we have seen that although KernelExplainer is model-agnostic, it can be very slow. So, I think that is one of the main motivations behind using LinearExplainer to explain a linear model with independent features and even consider feature correlation. In this section, we will discuss applying the LinearExplainer method in practice. The detailed notebook tutorial is available at `https://github.com/PacktPublishing/Applied-Machine-Learning-Explainability-Techniques/blob/main/Chapter07/LinearExplainers.ipynb`. We have used the same *Red Wine Quality dataset* as used for the tutorial discussed in *Chapter 6, Model Interpretability Using SHAP*. You can refer to the same tutorial to learn more about the dataset as we will only focus on the LinearExplainer application part in this section.

Application of LinearExplainer in SHAP

For this example, we have actually trained a linear regression model on the dataset. Similar to the other explainers, we can apply LinearExplainer using a few lines of code:

```
explainer = shap.LinearExplainer(model, x_train, feature_
dependence="independent")
shap_values = explainer.shap_values(x_test)
```

To explain the trained linear model, the same visualization methods covered for TreeExplainer and KernelExplainer are applicable. *Figure 7.10* shows the summary plot, feature dependence plot, and force plots used to explain the trained linear model:

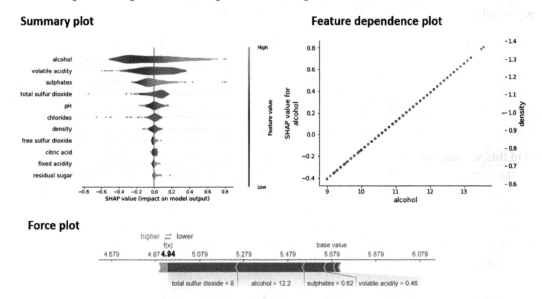

Figure 7.10 – Summary plot, feature dependence plot, and force plots obtained after using SHAP LinearExplainer

We can obtain the visualization plots shown in *Figure 7.10* using the following lines of code:

```
shap.summary_plot(shap_values, x_test, plot_type='violin',
show=False)
shap.force_plot(explainer.expected_value, shap_values[1], x_
test.iloc[0,:])
shap.dependence_plot("alcohol", shap_values, x_test,
show=False)
```

I will recommend that you explore other visualization methods or even create your custom visualizations using the SHAP values generated by the LinearExplainer. Next, we will discuss applying SHAP to transformer models trained on text data.

Explaining transformers using SHAP

In this chapter, so far we have seen examples of various SHAP explainers used to explain different types of models trained on structured and image datasets. Now, we will cover approaches to explain complicated models trained on text data. For text data, getting high accuracy with models trained on conventional **Natural Language Processing (NLP)** methods is always challenging. This is because extracting contextual information in sequential text data is always difficult using the classical approaches.

However, with the invention of the **Transformer** deep learning architecture (`https://blogs.nvidia.com/blog/2022/03/25/what-is-a-transformer-model/`), which is based on an **attention mechanism**, obtaining higher accuracy models trained on text data became much easier. However, transformer models are extremely complicated and it can be really difficult to interpret the workings of such models. Fortunately, being model-agnostic, SHAP can be applied with transformer models as well.

So in this section, we will cover how SHAP can be applied with *text-based, pre-trained transformer models from Hugging Face* (`https://github.com/huggingface/transformers`) used for different applications. The complete tutorial can be accessed from `https://github.com/PacktPublishing/Applied-Machine-Learning-Explainability-Techniques/blob/main/Chapter07/Explaining_Transformers.ipynb`. Now, let's see the first example of explaining transformer-based sentiment analysis models using SHAP.

Explaining transformer-based sentiment analysis models

Hugging Face (`https://huggingface.co/`) provides state-of-the-art pre-trained models trained on a large amount of data. Hence, using pre-trained models from Hugging Face, AI developers can actually focus on the problem solving and application part, rather than spending too much time on tasks such as data collection, processing, and model training. The first example that we will discuss will be for doing sentiment analysis from text data. Before we proceed with the application, please ensure that you have installed the `transformers` Python module using the `pip` installer:

```
!pip install --upgrade transformers
```

You can confirm the successful installation of the transformers framework by importing the module:

```
import transformers
```

Now, let's load a sentiment analysis pre-trained model:

```
model = transformers.pipeline('sentiment-analysis', return_all_
scores=True)
```

We can take any sentence as an input and we will apply the model to check whether it has positive or negative sentiment. So, we will use the sentence "Hugging Face transformers are absolutely brilliant!" as our inference data:

```
text_data = "Hugging Face transformers are absolutely
brilliant!"
model(text_data)[0]
```

The model will predict the probability of the inference data being positive and negative:

```
[{'label': 'NEGATIVE', 'score': 0.00013269631017465144},
 {'label': 'POSITIVE', 'score': 0.99986732006073}]
```

With a very high probability (99.99%), the model has predicted the sentence to be positive, which is a correct prediction. Now, let's apply SHAP to explain the model prediction:

```
explainer = shap.Explainer(model)
shap_values = explainer([text_data])
```

Once the SHAP values are successfully computed, we can apply SHAP text plot visualization and bar plot visualization to highlight words that are positively and negatively contributing to the model's prediction:

```
shap.plots.text(shap_values[0,:,'POSITIVE'])
shap.plots.bar(shap_values[0,:,'POSITIVE'])
```

Figure 7.11 shows us the SHAP text plots, which look similar to force plots:

Figure 7.11 – Explaining transformer models trained on text data using SHAP text plots

As we can see from *Figure 7.11*, the words highlighted in red, such as *brilliant, absolutely,* and *Hugging,* make a positive contribution and increase the model prediction score, whereas the other words are lowering the model prediction and thus make a negative contribution to the model's prediction.

The same inference can also be drawn from the SHAP bar plot shown in the following figure:

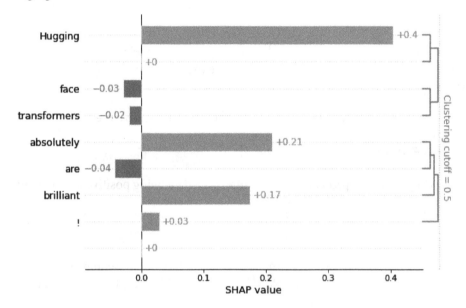

Figure 7.12 – SHAP bar plot used to explain a transformer model trained on text data

I find it easier to interpret bar plots, which clearly show the positive or negative impact of each word present in the sentence, as shown in *Figure 7.12*.

Next, let's explore another example, in which a transformer-based multi-class classification model is trained on text data.

Explaining a multi-class prediction transformer model using SHAP

In the previous example, we applied SHAP to explain a text-based binary classification model. Now, let's apply SHAP to explain a pre-trained transformer model used for detecting one of the following six emotions: *sadness, joy, love, anger, fear,* and *surprise.*

Let's load the pre-trained transformer model for emotion detection:

```
tokenizer = transformers.AutoTokenizer.from_
pretrained("nateraw/bert-base-uncased-emotion", use_fast=True)
model = transformers.AutoModelForSequenceClassification.from_
pretrained("nateraw/bert-base-uncased-emotion").cuda()


# build a pipeline object to do predictions
pipeline = transformers.pipeline("text-classification",
model=model, tokenizer=tokenizer, device=0, return_all_
scores=True)
```

Now, let's use the same inference data as in the previous example and compute the SHAP values using SHAP:

```
explainer = shap.Explainer(pipeline)
shap_values = explainer([text_data])
```

We can then use the SHAP text plot to highlight words that make a positive or negative contribution to each of the six possible outcomes:

```
shap.plots.text(shap_values[0])
```

Figure 7.13 illustrates the output of the SHAP text plot obtained from the previous line of code:

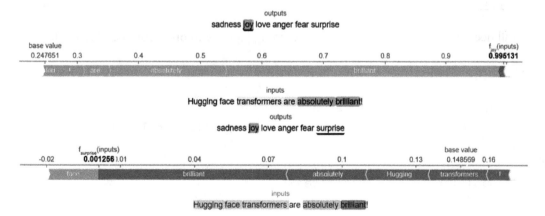

Figure 7.13 – Interactive SHAP text plot highlighting the words that make a positive and negative contribution

SHAP text plots are interactive. As we can see from *Figure 7.13*, this highlights the model's predicted outcome in red along with the words that make a positive and negative contribution to the model's decision. We can also click on other possible outcomes and visualize the influence of each word on the model's prediction. For example, if we click on *surprise* instead of *joy*, we will see that all the words other than the word *face* are highlighted in blue, as these words are contributing negatively to that specific outcome. Personally, I found this approach of explaining transformer models trained on text data using SHAP to be really interesting and efficient! Next, let's cover another interesting use case of applying SHAP to explain NLP zero-shot learning models.

Explaining zero-shot learning models using SHAP

Zero-shot learning is one of the most fascinating concepts in NLP, which involves applying models on inference data for predicting any custom category that is not used during the training process without the need for fine-tuning. You can find more information about zero-shot learning in this reference link: https://aixplain. com/2021/09/23/zero-shot-learning-in-natural-language- processing/. Applying SHAP to zero-shot learning models is also very straightforward.

First, we will need to load the pre-trained transformer models:

```
model = AutoModelForSequenceClassification.from_
pretrained("valhalla/distilbart-mnli-12-3")
tokenizer = AutoTokenizer.from_pretrained("valhalla/distilbart-
mnli-12-3")
```

We will need to create a custom zero-shot learning pipeline in order for SHAP to work:

```
class ZSLPipeline(ZeroShotClassificationPipeline):
    # Overwrite the __call__ method
    def __call__(self, *args):
        out = super().__call__(args[0], self.set_labels)[0]


        return [[{"label":x[0], "score": x[1]}  for x in
zip(out["labels"], out["scores"])]]

    def set_labels(self, labels: Union[str,List[str]]):
        self.set_labels = labels
```

We will then need to define the custom labels and the inference text data and configure the new labels for the zero-shot learning model. For this example, we have selected the text `"I love playing cricket!"` as our inference data and we want our zero-shot learning model to predict whether the inference text data belongs to the `insect`, `sports`, or `animal` class:

```
text = ["I love playing cricket!"]
labels = ["insect", "sports", "animal"]
model.config.label2id.update({v:k for k,v in
enumerate(labels)})
model.config.id2label.update({k:v for k,v in
enumerate(labels)})
pipe = ZSLPipeline(model=model, tokenizer=tokenizer, return_
all_scores=True)
pipe.set_labels(labels)
```

Once the process of setting up the zero-shot learning model is ready, we can easily apply SHAP for the model explainability:

```
explainer = shap.Explainer(pipe)
shap_values = explainer(text)
```

After the SHAP values have been computed successfully, we can use text plots or bar plots for the model explainability:

```
shap.plots.text(shap_values)
shap.plots.bar(shap_values[0,:,'sports'])
```

Figure 7.14 shows the bar plot visualization obtained as an output:

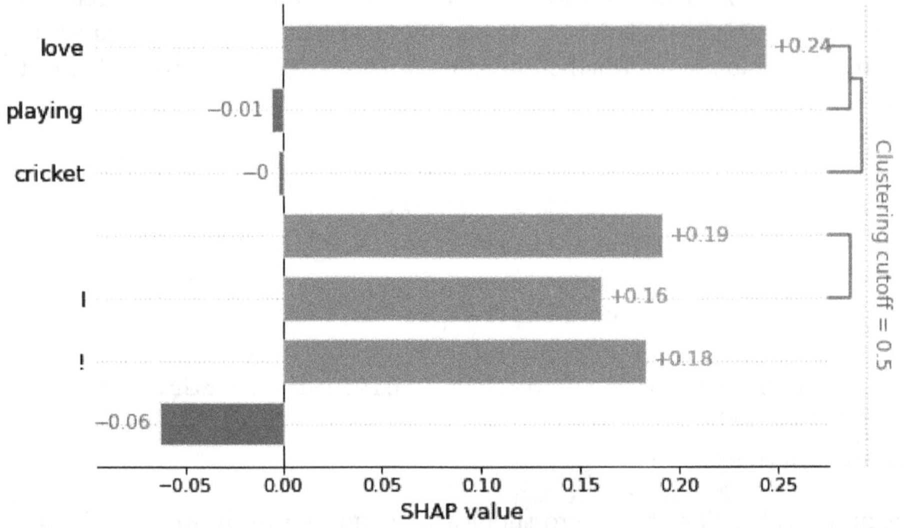

Figure 7.14 – Bar plot visualization for the outcome of the predicted sport
for the zero-shot learning model

The inference text sentence – `"I love playing cricket!"` used for this example was indeed related to the `Sports` class, which was correctly predicted by the model. However, cricket is not only a sport, but an insect as well. When the phrase `playing cricket` is used, collectively it indicates that we are talking about a sport. So, these words should make a positive contribution to the model's prediction. Unfortunately, from *Figure 7.14*, we can see that both the words `playing` and `cricket` are negatively contributing with negative SHAP values. This gives us an indication that even though the model prediction is correct, this is not a very good model as the model is relying heavily on the word `love` instead of the words `cricket` or `playing`. This is a classic example that highlights the need to make **Explainable AI (XAI)** a mandatory part of the AI process cycle and we should not blindly trust models even if the model prediction is correct.

We have now arrived at the end of this chapter, and I will summarize the important topics that we have discussed in this chapter.

Summary

After reading this chapter, you have received some practical exposure to using SHAP with tabular structured data as well as unstructured data such as images and texts. We have discussed the different explainers available in SHAP for both model-specific and model-agnostic explainability. We have applied SHAP to explain linear models, tree ensemble models, convolution neural network models, and even transformer models in this chapter. Using SHAP, we can explain different types of models trained on different types of data. I highly recommend trying out the end-to-end tutorials provided in the GitHub code repository and exploring things in more depth to acquire deeper practical knowledge.

In the next chapter, we will discuss another interesting topic of concept activation vectors and explore the practical part of applying the **Testing with Concept Activation Vectors (TCAV)** framework from Google AI for explaining models with human-friendly concepts.

References

Please refer to the following resources to gain additional information:

- *Shapley, Lloyd S. "A value for n-person games." Contributions to the Theory of Games 2.28 (1953)*: https://doi.org/10.1515/9781400881970-018

- *Red Wine Quality Dataset – Kaggle*: https://www.kaggle.com/uciml/red-wine-quality-cortez-et-al-2009

- *SHAP GitHub Project*: https://github.com/slundberg/shap

- *SHAP Documentation*: https://shap.readthedocs.io/en/latest/index.html

- *Hugging Face Models*: https://huggingface.co/

- *Zero-Shot Learning*: https://en.wikipedia.org/wiki/Zero-shot_learning

8
Human-Friendly Explanations with TCAV

In the previous few chapters, we have extensively discussed **LIME** and **SHAP**. You have also seen the practical aspect of applying the Python frameworks of LIME and SHAP to explain black-box models. One major limitation of both frameworks is that the method of explanation is not extremely consistent and intuitive with how non-technical end users would explain an observation. For example, if you have an image of a glass filled with Coke and use LIME and SHAP to explain a black-box model used to correctly classify the image as Coke, both LIME and SHAP would highlight regions of the image that lead to the correct prediction by the trained model. But if you ask a non-technical user to describe the image, the user would classify the image as Coke due to the presence of a dark-colored carbonated liquid in a glass that resembles a Cola drink. In other words, human beings tend to relate any observation with known *concepts* to explain it.

Testing with Concept Activation Vector (TCAV) from *Google AI* also follows a similar approach in terms of explaining model predictions with known *human concepts*. So, in this chapter, we will cover how TCAV can be used to provide concept-based human-friendly explanations. Unlike LIME and SHAP, TCAV works beyond *feature attribution* and refers to concepts such as *color, gender, race, shape, any known object,* or an *abstract idea* to explain model predictions. In this chapter, we will discuss the workings of the TCAV algorithm. I will cover some of the advantages and disadvantages of the framework. We will also discuss using this framework for practical problem-solving. In *Chapter 2, Model Explainability Methods,* under *Representation-based explanation,* you did get some exposure to TCAV, but in this chapter, we will cover the following topics:

- Understanding TCAV intuitively
- Exploring the practical applications of TCAV
- Advantages and limitations
- Potential applications of concept-based explanations

It's time to get started now!

Technical requirements

This code tutorial and the requisite resources can be downloaded or cloned from the GitHub repository for this chapter at `https://github.com/PacktPublishing/Applied-Machine-Learning-Explainability-Techniques/tree/main/Chapter08`. Similar to other chapters, Python and Jupyter notebooks are used to implement the practical application of the theoretical concepts covered in this chapter. However, I would recommend you run the notebooks only after you have gone through this chapter for a better understanding.

Understanding TCAV intuitively

The idea of TCAV was first introduced by *Kim et al.* in their work – *Interpretability beyond Feature Attribution: Quantitative Testing with Concept Activation Vectors (TCAV)* (`https://arxiv.org/pdf/1711.11279.pdf`). The framework was designed to provide interpretability beyond feature attribution, particularly for deep learning models that rely on low-level transformed features that are not human-interpretable. TCAV aims to explain the opaque internal state of the deep learning model using abstract, high-level, human-friendly concepts. In this section, I will present you with an intuitive understanding of TCAV and explain how it works to provide human-friendly explanations.

What is TCAV?

So far, we have covered many methods and frameworks to explain ML models through feature-based approaches. But it might occur to you that since most ML models operate on low-level features, the feature-based explanation approaches might highlight features that are not human-interpretable. For example, for explaining image classifiers, pixel intensity values or pixels coordinates in an image might not be useful for end users without any technical background in data science and ML. So, these features are not user-friendly. Moreover, feature-based explanations are always restricted by the selection of features and the number of features present in the dataset. Out of all the features selected by the feature-based explanation methods, end users might be interested in a particular feature that is not picked by the algorithm.

So, instead of this approach, concept-based approaches provide a much wider abstraction that is human-friendly and more relevant as interpretability is provided in terms of the importance of high-level concepts. So, **TCAV** is a model interpretability framework from Google AI that implements the idea of a concept-based explanation method in practice. The algorithm depends on **Concept Activation Vectors (CAV)**, which provide an interpretation of the internal state of ML models using human-friendly concepts. In a more technical sense, TCAV uses directional derivatives to quantify the importance of human-friendly, high-level concepts for model predictions. For example, while describing hairstyles, concepts such as *curly hair*, *straight hair*, or *hair color* can be used by TCAV. These user-defined concepts are not the input features of the dataset that are used by the algorithm during the training process.

The following figure illustrates the key question addressed by TCAV:

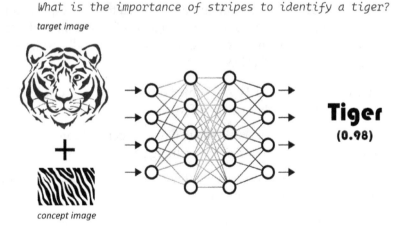

Figure 8.1 – TCAV helps us to address the key question of concept importance of a user-defined concept for image classification by a neural network

In the next section, let's try to understand the idea of model explanation using abstract concepts.

Explaining with abstract concepts

By now, you may have an intuitive understanding of the method of providing explanations with abstract concepts. But why do you think this is an effective approach? Let's take another example. Suppose you are working on building a deep learning-based image classifier for detecting doctors from images. After applying TCAV, let's say that you have found out that the *concept importance* of the concept *white male* is maximum, followed by *stethoscope* and *white coat*. The concept importance of *stethoscope* and *white coat* is expected, but the high concept importance of *white male* indicates a biased dataset. Hence, TCAV can help to evaluate **fairness** in trained models.

Essentially, the goal of CAVs is to estimate the importance of a concept (such as color, gender, and race) for the prediction of a trained model, even though the *concepts* were not used during the model training process. This is because TCAV learns *concepts* from a few example samples. For example, in order to learn a *gender* concept, TCAV needs a few data instances that have a *male* concept and a few *non-male* examples. Hence, TCAV can quantitatively estimate the trained model's sensitivity to a particular *concept* for that class. For generating explanations, TCAV perturbs data points toward a *concept* that is relatable to humans, and so it is a type of **global perturbation method**. Next, let's try to learn the main objectives of TCAV.

Goals of TCAV

I found the approach of TCAV to be very unique as compared to other explanation methods. One of the main reasons is because the developers of this framework established clear goals that resonate with my own understanding of human-friendly explanations. The following are the established goals of TCAV:

- **Accessibility**: The developers of TCAV wanted this approach to be accessible to any end user, irrespective of their knowledge of ML or data science.

- **Customization**: The framework can adapt to any user-defined concept. This is not limited to concepts considered during the training process.

- **Plug-in readiness**: The developers wanted this approach to work without the need to retrain or fine-tune trained ML models.

- **Global interpretability**: TCAV can interpret the entire class or multiple samples of the dataset with a single quantitative measure. It is not restricted to the local explainability of data instances.

Now that we have an idea of what can be achieved using TCAV, let's discuss the general approach to how TCAV works.

Approach of TCAV

In this section, we will cover the workings of TCAV in more depth. The overall workings of this algorithm can be summarized in the following methods:

- Applying directional derivatives to quantitatively estimate the sensitivity of predictions of trained ML models for various user-defined concepts.

- Computing the final quantitative explanation, which is termed **TCAVq measure**, without any model re-training or fine-tuning. This measure is the relative importance of each concept to each model prediction class.

Now, I will try to further simplify the approach of TCAV without using too many mathematical notions. Let's assume we have a model for identifying zebras from images. To apply TCAV, the following approach can be taken:

1. **Defining a concept of interest**: The very first step is to consider the concepts of interest. For our zebra classifier, either we can have a given set of examples that represent the concept (such as black stripes are important in identifying a zebra) or we can have an independent dataset with the concepts labeled. The major benefit of this step is that it does not limit the algorithm from using features used by the model. Even non-technical users or domain experts can define the concepts based on their existing knowledge.

2. **Learning concept activation vectors**: The algorithm tries to learn a vector in the space of activation of the layers by training a linear classifier to differentiate between activations generated by a concept's instances and instances present in any layer. So, a **CAV** is defined as the normal projection to a hyperplane that separates instances with a concept and instances without a concept in the model's activation. For our zebra classifier, CAVs help to distinguish representations that denote *black stripes* and representations that do not denote *black stripes*.

3. **Estimating directional derivatives**: Directional derivatives are used to quantify the sensitivity of a model prediction toward a concept. So, for our zebra classifier, directional directives help us to measure the importance of the *black stripes* representation in predicting zebras. Unlike saliency maps, which use per-pixel saliency, directional derivatives are computed on the entire dataset or a set of inputs but for a specific concept. This helps to give a global perspective for the explanation.

4. **Estimating the TCAV score**: To quantify the concept importance of a particular class, the TCAV score (**TCAVq**) is calculated. This metric helps to measure the positive or negative influence of a defined concept on a particular activation layer of a model.

5. **CAV validation**: CAV can be produced from randomly selected data. But unfortunately, this might not produce meaningful concepts. So, in order to improve the generated concepts, TCAV runs multiple iterations for finding concepts from different batches of data, instead of training CAV once, on a single batch of data. Then, a **statistical significance test** is performed using *two-side t-test* for selecting the statistically significant concepts. Necessary corrections, such as the *Bonferroni correction*, are also performed to control the false discovery rate.

Thus, we have covered the intuitive workings of the TCAV algorithm. Next, let's cover how TCAV can actually be implemented in practice.

Exploring the practical applications of TCAV

In this section, we will explore the practical applications of TCAV for explaining pre-trained image explainers with concept importance. The entire notebook tutorial is available in the code repository of this chapter at `https://github.com/ PacktPublishing/Applied-Machine-Learning-Explainability- Techniques/blob/main/Chapter08/Intro_to_TCAV.ipynb`. This tutorial is presented based on the notebook provided in the original GitHub project repository of TCAV `https://github.com/tensorflow/tcav`. I recommend that you all refer to the main project repository of TCAV since the credit for implementation should go to the developers and contributors of TCAV.

In this tutorial, we will cover how to apply TCAV to validate the concept importance of the concept of *stripes* as compared to the *honeycomb* pattern for identifying *tigers*. The following images illustrate the flow of the approach used by TCAV to ascertain concept importance using a simple visualization:

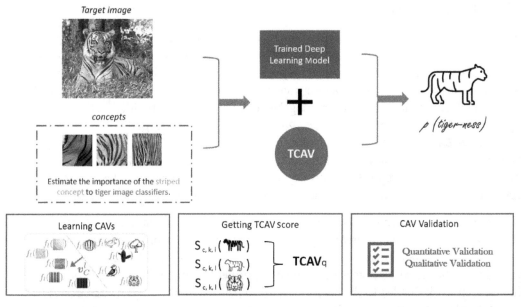

Figure 8.2 – Using TCAV to estimate the concept importance of stripes in a tiger image classifier

Let's begin by setting up our Jupyter notebook.

Getting started

Similar to the other tutorial examples covered in the previous chapters, to install the necessary Python modules required to run the notebook, you can use the `pip install` command in a Jupyter notebook:

```
!pip install --upgrade pandas numpy matplotlib tensorflow tcav
```

You can import all the modules to validate the successful installation of these frameworks:

```
import tensorflow as tf
import tcav
```

Next, let's take a look at the data that we'll be working with.

About the data

I felt that the data preparation process, which is provided in the original project repository of TCAV, is slightly time-consuming. So, I have already prepared the necessary datasets, which you can refer to from this project repository. Since we will be validating the importance of the concept of *stripes* for images of *tigers*, we will need an image dataset for tigers. The data is collected from the ImageNet collection and is provided in the project repository at `https://github.com/PacktPublishing/Applied-Machine-Learning-Explainability-Techniques/tree/main/Chapter08/images/tiger`. The images are randomly curated and collected using the *data collection script* provided in the TCAV repository: `https://github.com/tensorflow/tcav/tree/master/tcav/tcav_examples/image_models/imagenet`.

In order to run TCAV, you would need to have the necessary *concept images, target class images,* and *random dataset images.* For this tutorial, I have prepared the concept images from the *Broden dataset* (`http://netdissect.csail.mit.edu/data/broden1_224.zip`), as suggested in the main project example. Please go through the research work that led to the creation of this dataset: `https://github.com/CSAILVision/NetDissect`. You can also explore the *Broden dataset texture images* provided at `https://github.com/PacktPublishing/Applied-Machine-Learning-Explainability-Techniques/tree/main/Chapter08/concepts/broden_concepts` to learn more. I recommend you to experiment with other concepts or other images and play around with TCAV-based concept importance!

> **Broden dataset**
>
> *David Bau*, Bolei Zhou*, Aditya Khosla, Aude Oliva, and Antonio Torralba.*
> *Network Dissection: Quantifying Interpretability of Deep Visual Representations.*
> *Computer Vision and Pattern Recognition (CVPR), 2017.*

As TCAV also requires some random datasets to ascertain the statistical significance of the concepts learned from target image examples, I have provided some sample random images in the project repository, thereby simplifying the running of the tutorial notebook! But as always, you should experiment with other random image examples as well. These random images are also collected using the image fetcher script provided in the main project: `https://github.com/tensorflow/tcav/blob/master/tcav/tcav_examples/image_models/imagenet/download_and_make_datasets.py`.

To proceed further, you need to define the variables for the target class and the concepts:

```
target = 'tiger'
concepts = ['honeycombed', 'striped']
```

You can also create the necessary paths and directories to store the generated activations and CAVs as mentioned in the notebook tutorial. Next, let's discuss the model used in this example.

Discussions about the deep learning model used

In this example, we will use a pre-trained deep learning model to highlight the fact that even though TCAV is considered to be a model-specific approach, as it is only applicable to neural networks, it does not make an assumption of the network architecture as such and can work well with most deep neural network models.

For this example, we will use the pre-trained GoogleNet model, `https://paperswithcode.com/method/googlenet`, based on the ImageNet dataset (`https://www.image-net.org/`). The model files are provided in the code repository: `https://github.com/PacktPublishing/Applied-Machine-Learning-Explainability-Techniques/tree/main/Chapter08/models/inception5h`. You can load the trained model using the following lines of code:

```
model_to_run = 'GoogleNet'
sess = utils.create_session()
GRAPH_PATH = "models/inception5h/tensorflow_inception_graph.pb"
LABEL_PATH = "models/inception5h/imagenet_comp_graph_label_
strings.txt"

trained_model = model.GoogleNetWrapper_public(sess,
                                              GRAPH_PATH,
                                              LABEL_PATH)
```

The model wrapper is actually used to get the internal state and tensors of the trained model. Concept importance is actually computed based on the internal neuron activations and hence, this model wrapper is important. For more details about the workings of the internal API, please refer to the following link: `https://github.com/tensorflow/tcav/blob/master/Run_TCAV_on_colab.ipynb`.

Next, we would need to generate the concept activations using the `ImageActivationGenerator` method:

```
act_generator = act_gen.ImageActivationGenerator(
    trained_model, source_dir, activation_dir,
    max_examples=100)
```

Next, we will explore model explainability using TCAV.

Model explainability using TCAV

As discussed before, TCAV is currently used to explain neural networks and the inner layers of a neural network. So, it is not model-agnostic, but rather a model-centric explainability method. This requires us to define the bottleneck layer of the network:

```
bottlenecks = [ 'mixed4c']
```

The next step will be to apply the TCAV algorithm to create the concept activation vectors. The process also includes performing statistical significance testing using two side t-test between the concept importance of the target class and the random samples:

```
num_random_exp= 15


mytcav = tcav.TCAV(sess,
                   target,
                   concepts,
                   bottlenecks,
                   act_generator,
                   [0.1],
                   cav_dir=cav_dir,
                   num_random_exp=num_random_exp)
```

The original experiment mentioned in the TCAV paper, https://arxiv.org/abs/1711.11279, mentioned using at least 500 random experiments to identify the statistically significant concepts. But for the sake of simplicity, and to achieve faster results, we are using 15 random experiments. You can experiment with more random experiments as well.

Finally, we can get the results and visualize the concept importance:

```
results = mytcav.run(run_parallel=False)
utils_plot.plot_results(results,
                        num_random_exp=num_random_exp)
```

This will generate the following plot that helps us to compare concept importance:

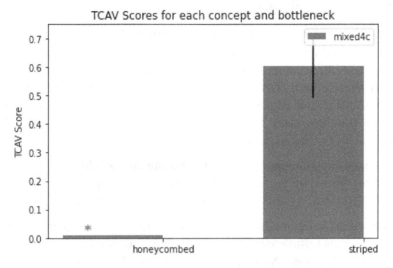

Figure 8.3 – TCAV concept importance of the concepts of striped and honeycombed
for identifying tiger images

As you can observe from *Figure 8.3*, the *striped* concept has significantly higher concept importance than the *honeycombed* concept for identifying *tigers*.

Now that we have covered the practical application part, let me give you a similar challenge as an exercise. Can you now use the ImageNet dataset and ascertain the importance of the concept of *water* to *ships* and of *clouds* or *sky* to *airplanes*? This will help you understand this concept in more depth and give you more confidence to apply TCAV. Next, we will discuss some advantages and limitations of TCAV.

Advantages and limitations

In the previous section, we covered the practical aspects of TCAV. TCAV is indeed a very interesting and novel approach to explaining complex deep learning models. Although it has many advantages, unfortunately, I did find some limitations in terms of the current framework that can definitely be improved in the revised version.

Advantages

Let's discuss the following advantages first:

- As you have previously seen with the LIME framework in *Chapter 4, LIME for Model Interpretability* (which generates explanations using a **global perturbation method**), there can be contradicting explanations for two data instances for the same class. Even though TCAV is also a type of global perturbation method, unlike LIME, TCAV-generated explanations are not only true for a single data instance but also true for the entire class. This is a major advantage of TCAV over LIME, which increases the user's trust in the explanation method.

- Concept-based explanations are closer to how humans would explain an unknown observation, rather than feature-based explanations as adopted in LIME and SHAP. So, TCAV-generated explanations are indeed more human-friendly.

- Feature-based explanations are limited to the features used in the model. To introduce any new feature for model explainability, we would need to re-train the model, whereas a concept-based explanation is more flexible and is not limited to features used during model training. To introduce a new concept, we do not need to retrain the model. Also, for introducing the concepts, you don't have to know anything about ML. You would just have to make the necessary datasets to generate concepts.

- Model explainability is not the only benefit of TCAV. TCAV can help to detect issues during the training process, such as **imbalanced datasets** leading to *bias in the dataset vis-à-vis the majority class*. In fact, concept importance can be used as a *metric* to compare models. For example, suppose you are using a *VGG19* model and a *ResNet50* model. Let's say both these models have similar accuracy and model performance, yet concept importance for a user-defined concept is much higher for the VGG19 model as compared to the ResNet50 model. In such a case, it is better to use the VGG19 model as compared to ResNet50. Hence, TCAV can be used to improve the model training process.

These are some of the distinct advantages of TCAV, which makes it more human-friendly than LIME and SHAP. Next, let's discuss some known limitations of TCAV.

Limitations

The following are some of the known disadvantages of the TCAV approach:

- Currently, the approach of concept-based explanation using TCAV is limited to just neural networks. In order to increase its adoption, TCAV would need an implementation that can work with *classical machine learning algorithms* such as *Decision Trees, Support Vector Machines*, and *Ensemble Learning algorithms*. Both LIME and SHAP can be applied with classical ML algorithms to solve standard ML problems and that is probably why LIME and SHAP have more adoption. Similarly, with text data, too, TCAV has very limited applications.

  TCAV is highly prone to *data drift, adversarial effects*, and *other data quality issues* discussed in *Chapter 3, Data-Centric Approaches*. If you are using TCAV, you would need to ensure that training data, inference data, and even concept data have similar statistical properties. Otherwise, the concepts generated can become affected due to noise or data impurity issues:

- *Guillaume Alain* and *Yoshua Bengio*, in their paper *Understanding intermediate layers using linear classifier probes* (`https://arxiv.org/abs/1610.01644`), have expressed some concern about applying TCAV to shallower neural networks. Many similar research papers have suggested that concepts in deeper layers are more separable as compared to concepts in shallower networks and, hence, the use of TCAV is limited to mostly deep neural networks.

- Preparing a concept dataset can be a challenging and expensive task. Although you don't need ML knowledge to prepare a concept dataset, still, in practice, you do not expect any common end user to spend time creating an annotated concept dataset for any customized user-defined concept.

- I felt that the TCAV Python framework would require further improvements before being used in any production-level system. In my opinion, at the time of writing this chapter, this framework would need to mature further so that it can be used easily with any production-level ML system.

I think all these limitations can indeed be solved to make TCAV a much more robust framework that is widely adopted. If you are interested, you can also reach out to authors and developers of the TCAV framework and contribute to the open source community! In the next section, let's discuss some potential applications of concept-based explanations.

Potential applications of concept-based explanations

I do see great potential for concept-based explanations such as TCAV! In this section, you will get exposure to some potential applications of concept-based explanations that can be important research topics for the entire AI community, which are as follows:

- **Estimation of transparency and fairness in AI**: Most regulatory concerns for black-box AI models are related to concepts such as gender, color, and race. Concept-based explanations can actually help to estimate whether an AI algorithm is fair in terms of these abstract concepts. The detection of bias for AI models can actually improve its transparency and help to address certain regulatory concerns. For example, in terms of doctors using deep learning models, TCAV can be used to detect whether the model is biased toward a specific gender, color, or race as ideally, these concepts are not important as regards the model's decision. High concept importance for these concepts indicates the presence of bias. *Figure 8.4* illustrates an example where TCAV is used to detect model bias.

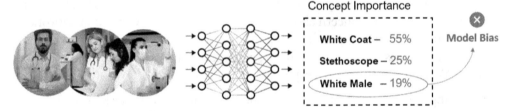

Figure 8.4 – TCAV can be used to detect model bias based on concept importance

- **Detection of adversarial attacks with CAV**: If you go through the appendix of the TCAV research paper (https://arxiv.org/pdf/1711.11279.pdf), the authors have mentioned that the concept importance of actual samples and adversarial samples are quite different. This means that if an image gets impacted by an adversarial attack, the concept importance would also change. So, CAVs can be a potential method in detecting adversarial attacks, as discussed in *Chapter 3, Data-Centric Approaches*.

- **Concept-based image clustering**: Using CAVs to cluster images based on similar concepts can be an interesting application. Deep learning-based image search engines are a common application in which clustering or similarity algorithms are applied to feature vectors to locate similar images. However, these are feature-based methods. Similarly, there is a potential to apply concept-based image clustering using CAVs.

- **Automated concept-based explanations (ACE)**: *Ghorbani, Amirata, James Wexler, James Zou, and Been Kim*, in their research work – *Towards automatic concept-based explanations*, mentioned an automated version of TCAV that goes through the training images and automatically discovers prominent concepts. This is an interesting work, as I think it can have an important application in identifying incorrectly labeled training data. In industrial applications, getting a perfectly labeled curated dataset is extremely challenging. This problem can be solved to a great extent using ACE.

- **Concept-based Counterfactual Explanation**: In *Chapter 2*, *Model Explainability Methods*, we discussed **counterfactual explanation (CFE)** as a mechanism for generating actionable insights by suggesting changes to the input features that can change the overall outcome. However, CFE is a feature-based explanation method. It would be a really interesting topic of research to have a concept-based counterfactual explanation, which is one step closer to human-friendly explanations.

For example, if we say that it is going to *rain* today although there is a *clear sky* now, we usually add a further explanation that suggests that the *clear sky* can be covered with *clouds*, which increases the probability of rainfall. In other words, a *clear sky* is a concept related to a *sunny* day, while a *cloudy sky* is a concept related to *rainfall*. This example suggests that the forecast can be flipped if the concept describing the situation is also flipped. Hence, this is the idea of the concept-based counterfactual. The idea is not very far-fetched as **concept bottleneck models (CBMs)** presented in the research work by *Koh et al.*, in https://arxiv.org/abs/2007.04612, can implement a similar idea of generated concept-based counterfactuals by manipulating the neuron action of the bottleneck layer.

Figure 8.5 illustrates an example of using a concept-based counterfactual example. There is no existing algorithm or framework that can help us achieve this, yet this can be a useful application of concept-based approaches in computer vision.

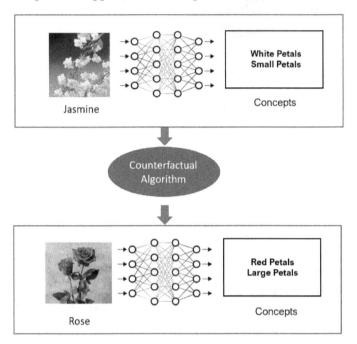

Figure 8.5 – An illustration of the idea of concept-based counterfactual examples

I feel this is a wide-open research field and the potential to come up with game-changing applications using concept-based explanations is immense. I do sincerely hope that more and more researchers and AI developers start working on this area to make significant progress in the coming years! Thus, we have arrived at the end of this chapter. Let me now summarize what has been covered in this chapter in the next section.

Summary

This chapter covers the concepts of TCAV, a novel approach, and a framework developed by Google AI. You have received a conceptual understanding of TCAV, practical exposure to applying the Python TCAV framework, learned about some key advantages and limitations of TCAV, and finally, I presented some interesting ideas regarding potential research problems that can be solved using concept-based explanations.

In the next chapter, we will explore other popular XAI frameworks and apply these frameworks to solving practical problems.

References

Please refer to the following resources to gain additional information:

- *Interpretability Beyond Feature Attribution: Quantitative Testing with Concept Activation Vectors (TCAV)*: `https://arxiv.org/pdf/1711.11279.pdf`

- *TCAV Python framework* - `https://github.com/tensorflow/tcav`

- *Koh et al. "Concept Bottleneck Models"*: `https://arxiv.org/abs/2007.04612`

- *Guillaume Alain and Yoshua Bengio, "Understanding intermediate layers using linear classifier probes"*: `https://arxiv.org/abs/1610.01644`

- *Ghorbani, Amirata, James Wexler, James Zou and Been Kim, "Towards automatic concept-based explanations"*: `https://arxiv.org/abs/1902.03129`

- *Detecting Concepts, Chapter 10.3 Molnar, C. (2022). Interpretable Machine Learning: A Guide for Making Black Box Models Explainable (2nd ed.).*: `https://christophm.github.io/interpretable-ml-book/detecting-concepts.html`

9
Other Popular XAI Frameworks

In the previous chapter, we covered the **TCAV framework** from **Google AI**, which is used for producing *human-friendly concept-based explanations*. We also discussed the other widely used explanation frameworks: **LIME** and **SHAP**. However, LIME, SHAP, and even TCAV have certain limitations, which we discussed in earlier chapters. None of these frameworks covers all the four dimensions of explainability for non-technical end-users. Due to these known drawbacks, the search for a robust **Explainable AI (XAI)** framework is still on.

The journey toward finding a robust XAI framework and addressing the known limitations of the popular XAI modules has led to the discovery and development of many other robust frameworks trying to address different aspects of ML model explainability. In this chapter, we will cover these other popular XAI frameworks apart from LIME, SHAP and TCAV.

More specifically, we will discuss about the important features, and key advantages of each of these frameworks. We will also explore how you can apply each framework in practice. Covering everything about each framework is beyond the scope of this chapter. But in this chapter, you will learn the most important features and practical application of these frameworks. In this chapter, we will cover the following list of widely used XAI frameworks:

- DALEX
- Explainerdashboard
- InterpretML
- ALIBI
- DiCE
- ELI5
- H2O AutoML explainer

At the end of the chapter, I will also share a quick comparison guide comparing all these XAI frameworks to help you to decide on the framework depending on your problem. Now, let's begin!

Technical requirements

This code tutorial with necessary resources can be downloaded or cloned from the GitHub repository for this chapter: `https://github.com/PacktPublishing/Applied-Machine-Learning-Explainability-Techniques/tree/main/Chapter09`. Like the other chapters, the Python and Jupyter notebooks are used to implement the practical application of the theoretical concepts covered in this chapter. But I will recommend you run the notebooks only after you go through this chapter for a better understanding. Most of the datasets used in the tutorials are also provided in the code repository: `https://github.com/PacktPublishing/Applied-Machine-Learning-Explainability-Techniques/tree/main/Chapter09/datasets`.

DALEX

In the *Dimensions of explainability* section of *Chapter 1, Foundational Concepts of Explainability Techniques*, we discussed the four different dimensions of explainability – *data*, *model*, *outcome*, and *end user*. Most explainability frameworks such as LIME, SHAP, and TCAV provide model-centric explainability.

DALEX (moDel Agnostic Language for Exploration and eXplanation) is one of the very few widely used XAI frameworks that tries to address most of the dimensions of explainability. DALEX is model-agnostic and can provide some metadata about the underlying dataset to give some context to the explanation. This framework gives you insights into the model performance and model fairness, and it also provides global and local model explainability.

The developers of the DALEX framework wanted to comply with the following list of requirements, which they have defined in order to explain complex black-box algorithms:

- **Prediction's justifications**: According to the developers of DALEX, ML model users should be able to understand the variable or feature attributions of the final prediction.

- **Prediction's speculations**: Hypothesizing the what-if scenarios or understanding the sensitivity of particular features of a dataset to the model outcome are other factors considered by the developers of DALEX.

- **Prediction's validations**: For each predicted outcome of a model, the users should be able to verify the strength of the evidence that confirms a particular prediction of the model.

DALEX is designed to comply with the preceding requirements using the various explanation methods provided by the framework. *Figure 9.1* illustrates the model exploration stack of DALEX:

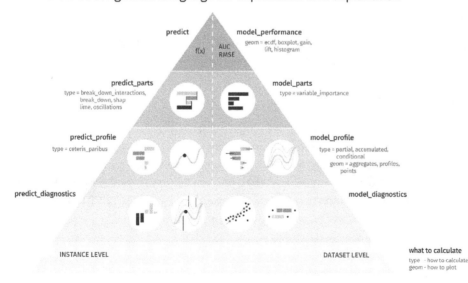

Figure 9.1 – The model exploration stack of DALEX

Next, I will walk you through an example of how to explore DALEX for explaining a black-box model in practice.

Setting up DALEX for model explainability

In this section, you will learn to setup DALEX in Python. Before starting the code walk-through, I would ask you to check the notebook at https://github.com/ PacktPublishing/Applied-Machine-Learning-Explainability-Techniques/blob/main/Chapter09/DALEX_example.ipynb. It contains the steps needed to understand the concept that we are going to now discuss in depth. I also recommend that you take a look at the GitHub project repository of DALEX at https://github.com/ModelOriented/DALEX/tree/master/python/ dalex in case you need additional details while executing the notebook.

The DALEX Python framework can be installed using the pip installer:

```
pip install dalex -U
```

If you want to use any additional features of DALEX that require an optional dependency, you can try the following command:

```
pip install dalex[full]
```

You can validate the successful installation of the package by importing it into the Jupyter notebooks using the following command:

```
import dalex as dx
```

Hopefully, your import should be successful; otherwise, if you get any errors, you will need to reinstall the framework or separately install its dependencies.

Discussions about the dataset

Next, let's briefly about the dataset that is being used for this example. For this example, I have used the *FIFA Club Position Prediction dataset* (https://www.kaggle.com/ datasets/adityabhattacharya/fifa-club-position-prediction-dataset) to predict the valuation of a football player in Euros, based on their skill and abilities. So, this is a regression problem that can be solved by regression ML models.

FIFA Club Position dataset citation

Bhattacharya A. (2022). Kaggle - FIFA Club Position Prediction
dataset: `https://www.kaggle.com/datasets/`
`adityabhattacharya/fifa-club-position-prediction-`
`dataset`

Similar to all other standard ML solution flows, we start with the data inspection process. The dataset can be loaded as a pandas DataFrame, and we can inspect the dimension of the dataset, the features that are present, and the data type of each feature. Additionally, we can perform any necessary data transformation steps such as dropping irrelevant features, checking for missing values, and data imputation to fill in missing values for relevant features. I recommend that you follow the necessary steps provided in the notebook, but feel free to include other additional steps and explore the dataset in more depth.

Training the model

For this example, I have used a random forest regressor algorithm to fit a model after dividing the data into the training set and the validation set. This can be done using the following lines of code:

```
x_train,x_valid,y_train,y_valid = train_test_split(
  df_train,labels,test_size=0.2,random_state=123)
model = RandomForestRegressor(
  n_estimators=790, min_samples_split = 3,
  random_state=123).fit(x_train, y_train)
```

We do minimum hyperparameter tuning to train the model as our objective is not to build a highly efficient model. Instead, our goal is to use this model as a black-box model and use DALEX to explain the model. So, let's proceed to the model explainability part using DALEX.

Model explainability using DALEX

DALEX is model-agnostic as it does not assume anything about the model and can work with any algorithm. So, it considers the model as a black box. Before exploring how to use DALEX in Python, let's discuss the following key advantages of this framework:

- **DALEX provides a uniform abstraction over different prediction models**: As an explainer, DALEX is quite robust and works well with different types of model frameworks such as scikit-learn, H2O, TensorFlow, and more. It can work with data provided in different formats such as a NumPy array or pandas DataFrame. It provides additional metadata about the data or the model, which makes it easier to develop an end-to-end model explainability pipeline in production.

- **DALEX has a robust API structure**: The concise API structure of DALEX ensures that a consistent grammar and coding structure is used for model analysis. Using just a few lines of code, we can apply the various explainability methods and explain any black-box model.

- **It can provide local explanations for an inference data instance**: Prediction-level explainability for a single inference data instance can be easily obtained during DALEX. There are different methods available in DALEX such as interactive breakdown plots, SHAP feature importance plots, and what-if analysis plots, which can be used for local explanations. We will cover these methods, in more detail, in the next section.

- **It can also provide global explanations while considering the entire dataset and the model**: Model-level global explanations can also be provided using DALEX partial dependence plots, accumulated dependence plots, global variable importance plots, and more.

- **Bias and fairness checks can be easily done using DALEX**: DALEX provides quantitative ways in which to measure model fairness and bias. Unlike DALEX, most of the XAI frameworks do not provide explicit methods to evaluate model fairness.

- **The DALEX ARENA platform can be used to build an interactive dashboard for better user engagement**: DALEX can be used to create an interactive web app platform that can be used to design a custom dashboard to show interactive visualizations for the different model explainability methods that are available in DALEX. I think this unique feature of DALEX gives you the opportunity to create better user engagement by providing a tailor-made dashboard to meet the specific end user's needs.

Considering all of these key benefits, let's now proceed with learning how to apply these features available in DALEX.

First, we need to create a DALEX model explainer object, which takes the trained model, data, and model type as input. This can be done using the following lines of code:

```
# Create DALEX Explainer object
explainer = dx.Explainer(model,
                         x_valid, y_valid,
                         model_type = 'regression',
                         label='Random Forest')
```

Once the explainer object has been created, it also provides additional metadata about the model, which is shown as follows.

```
Preparation of a new explainer is initiated

-> data               : 317 rows 73 cols
-> target variable    : Parameter 'y' was a pandas.Series. Converted to a numpy.ndarray.
-> target variable    : 317 values
-> model_class        : sklearn.ensemble._forest.RandomForestRegressor (default)
-> label              : Random Forest
-> predict function   : <function yhat_default at 0x0000024E0F2E55E8> will be used (default)
-> predict function   : Accepts pandas.DataFrame and numpy.ndarray.
-> predicted values   : min = 2.83e+05, mean = 1.35e+07, max = 1.07e+08
-> model type         : regression will be used
-> residual function  : difference between y and yhat (default)
-> residuals          : min = -4.17e+07, mean = -5.39e+05, max = 1.08e+07
-> model_info         : package sklearn
```

Figure 9.2 – The DALEX explainer metadata

This initial metadata is very useful for building automated pipelines for certain production-level systems. Next, let's explore some model-level explanations provided by DALEX.

Model-level explanations

Model-level explanations are global explanations produced by DALEX. The consider model performance and the overall impact of all the features considered during prediction. The performance of a model can be checked using a single line of code:

```
model_performance = explainer.model_performance("regression")
```

Depending upon the type of ML model, different model evaluation metrics can be applied. In this example, we are dealing with a regression problem, and hence, DALEX uses the metrics MSE, RMSE, R^2, MAE, and so on. For a classification problem, metrics such as accuracy, precision, recall, and more will be used. As covered in *Chapter 3, Data-Centric Approaches*, by evaluating the model performance, we get to estimate the *data forecastability* of the model, which gives us an indication of the degree of correctness of the predicted outcome.

DALEX provides methods such as global feature importance, **partial dependence plots** (**PDPs**), and accumulated dependency plots to analyze the feature-based explanations for model-level predictions. First, let's try out the variable of feature importance plots:

```
Var _ Importance = explainer.model _ parts(
   variable _ groups=variable _ groups, B=15, random _ state=123)
Var _ Importance.plot(max _ vars=10,
                      rounding _ function=np.rint,
                      digits=None,
                      vertical _ spacing=0.15,
                      title = 'Feature Importance')
```

This will produce the following plot:

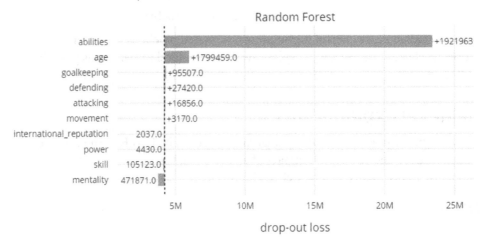

Figure 9.3 – A feature importance plot from DALEX for global feature-based explanations

In *Figure 9.3*, we can see that the trained model considers the abilities of the players, which comprise the overall rating of the player, the potential rating of the player, and other abilities such as pace, dribbling skill, strength, and stamina to be the most important factors for deciding the player's valuation.

Similar to feature importance, we can generate PDPs. Accumulated dependency plots can also be generated using the following few lines of code:

```
pdp = explainer.model_profile(type = 'partial', N=800)
pdp.plot(variables = ['age', 'potential'])
ald = explainer.model_profile(type = 'accumulated', N=800)
ald.plot(variables = ['age', 'movement_reactions'])
```

This will create the following plots for the aggregated profiles of the players:

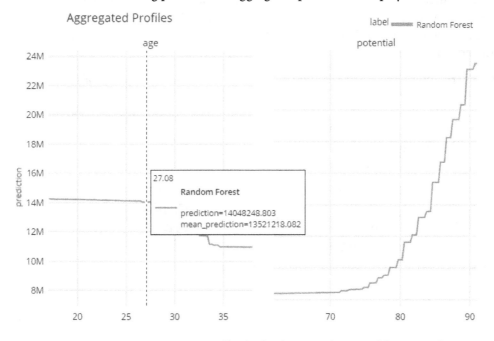

Figure 9.4 – A PDP aggregate profile plot for the age and potential features with predictions for model-level explanations

Figure 9.4 shows how the overall features of age and potential vary with the predicted valuation of football players. From the plot, we can understand that with an increase in the player's age, the predicted valuation decreases. Similarly, with an increase in a player's potential rating, the player's valuation increases. All of these observations are also quite consistent with the real-world observation for deciding a player's valuation. Next, let's see how to obtain a prediction-level explanation using DALEX.

Prediction-level explanations

DALEX can provide model-agnostic local or prediction-level explanations along with a global explanation. It uses techniques such as interactive breakdown profiles, SHAP feature importance values, and Ceteris Paribus profiles (what-if profiles) to explain model predictions at the individual data instance level. To understand the practical importance of these techniques, let's use these techniques for our use case to explain an ML model trained to predict the overall valuation of a football player. For our example, we will compare the prediction-level explanations of three players – Cristiano Ronaldo, Lionel Messi, and Jadon Sancho.

First, let's try out interactive breakdown plots. This can be done using the following lines of code:

```
prediction _ level = {'interactive _ breakdown':[], 'shap':[]}
ibd = explainer.predict _ parts(
   player, type='break _ down _ interactions', label=name)
prediction _ level['interactive _ breakdown'].append(ibd)
prediction _ level['interactive _ breakdown'][0].plot(
   prediction _ level['interactive _ breakdown'][1:3],
   rounding _ function=lambda x,
   digits: np.rint(x, digits).astype(np.int),
   digits=None,
   max _ vars=15)
```

This will generate the following interactive breakdown profile plot for each player:

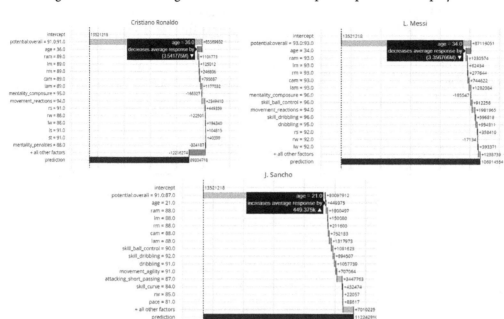

Figure 9.5 – An interactive breakdown plot from DALEX

Figure 9.5 shows the interactive breakdown plot comparing the model predictions for the three selected players. This plot illustrates the contribution of each feature to the final predicted value. The feature values that increase the model prediction value are shown in a different color than the features that decrease the prediction value. This plot shows the breakdown of the total predicted value with respect to each feature value of the data instance.

Now, all three players are world-class professional football players; however, Ronaldo and Messi are veteran players and living legends of the game, as compared to Sancho, who is a young talent. So, if you observe the plot, it shows that for Ronaldo and Messi, the age feature reduces the predicted value, while for Sancho, it slightly increases. It is quite interesting to observe that the model has been able to learn how increasing the age of football players can reduce their valuation. This observation is also consistent with the observation of domain experts who value younger players with higher potential to have a higher market value. Similar to breakdown plots, DALEX also provides SHAP feature importance plots to analyze the contribution of the features. This method gives similar information such as breakdown plots, but the feature importance is calculated based on SHAP values. It can be obtained using the following lines of code:

```
sh = explainer.predict_parts(player, type='shap', B=10,
                            label=name)
prediction_level['shap'].append(sh)
prediction_level['shap'][0].plot(
   prediction_level['shap'][1:3],
   rounding_function=lambda x,
   digits: np.rint(x, digits).astype(np.int),
   digits=None,
   max_vars=15)
```

Next, we will use *What-If* plots based on the **Ceteris Paribus profile** in DALEX. The Ceteris Paribus profile is similar to **Sensitivity Analysis**, which was covered in *Chapter 2, Model Explainability Methods*. It is based on the *Ceteris Paribus principle*, which means that when everything else remains unchanged, we can determine how a change in a particular feature will affect the model prediction. This process is often referred to as **What-If model analysis** or **Individual Conditional Expectations**. In terms of application, in our example, we can use this method to find out how the predicted valuation of Jadon Sancho might vary as he grows older or if his overall potential increases. We can find this out by using the following lines of code:

```
ceteris_paribus_profile = explainer.predict_profile(
    player,
    variables=['age', 'potential'],
    label=name) # variables to calculate

ceteris_paribus_profile.plot(size=3,
                              title= f"What If? {name}")
```

This will produce the following interactive what-if plot in DALEX:

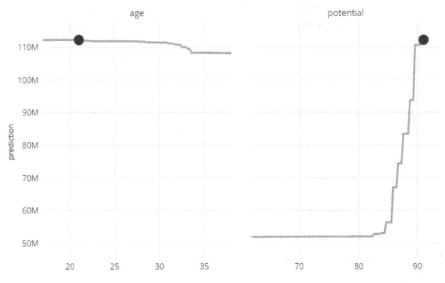

Figure 9.6 – An interactive what-if plot in DALEX

Figure 9.6 shows that for Jadon Sancho, the market valuation will start decreasing as he grows older; however, it can also increase with an increase in overall potential ratings. I would strongly recommend that you explore all of these prediction-level explanation options for the features and from the tutorial notebook provided in the project repository: `https://github.com/PacktPublishing/Applied-Machine-Learning-Explainability-Techniques/blob/main/Chapter09/DALEX_example.ipynb`. Next, we will use DALEX to evaluate model fairness.

Evaluating model fairness

A **Model Fairness** check is another important feature of DALEX. Although, model fairness and bias detection are more important to consider for classification problems relying on features related to gender, race, ethnicity, nationality, and other similar demographic features. However, we will apply this to regression models, too. For more details regarding model fairness checks using DALEX, please refer to `https://dalex.drwhy.ai/python-dalex-fairness.html`. Now, let's see whether our model is free from any bias and is fair!

We will create a **protected variable** and **privileged variable** for the fairness check. In fairness checks in ML, we try to ensure that a protected variable is free from any bias. If we anticipate any feature value or group to have any bias due to factors such as an imbalanced dataset, we can declare them as privileged variables. For our use case, we will perform a fairness check for three different sets of players based on their age.

All players less than 20 years are considered to be *youth* players, players between the ages of 20 and 30 are considered to be *developing* players, and players above 30 years are considered to be *developed* players. Now, let's do our fairness check using DALEX:

```
protected = np.where(x_valid.age < 30, np.where(x_valid.age <
20, 'youth', 'developing'), 'developed')
privileged = 'youth'
fairness = explainer.model_fairness(protected=protected,
                                    privileged=privileged)
fairness.fairness_check(epsilon = 0.7)
```

This is the outcome of the fairness checks:

```
No bias was detected! Conclusion: your model is fair in terms
of checked fairness criteria.
```

We can also check the quantitative evidence of the fairness checks and plot them to analyze further:

```
fairness.result
fairness.plot()
```

This will generate the following plot for analyzing model fairness checks using DALEX:

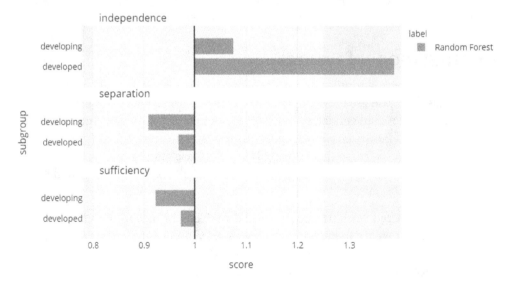

Figure 9.7 – A model fairness plot using DALEX

As shown in *Figure 9.7*, the model fairness using DALEX for regression models is done with respect to the metrics of independence, separation, and sufficiency. For classification models, these metrics could vary, but the API function usage is the same. Next, we will discuss the ARENA web-based tool for building interactive dashboards using DALEX.

Interactive dashboards using ARENA

Another interesting feature of DALEX is the ARENA dashboard platform to create an interactive web app that can be used to design a custom dashboard for keeping all the DALEX interactive visualizations that were obtained using different model explainability methods. This particular feature of DALEX gives us an opportunity to create better user engagement by creating a custom dashboard for a specific problem.

Before starting, we need to create a DALEX Arena dataset:

```
arena_dataset = df_test[:400].set_index('short_name')
```

Next, we need to create an `Arena` object and push the DALEX explainer object created from the black-box model that is being explained:

```
arena = dx.Arena()
# push DALEX explainer object
arena.push_model(explainer)
```

Following this, we just need to push the Arena dataset and start the server to make our Arena platform live:

```
# push whole test dataset (including target column)
arena.push_observations(arena_dataset)
# run server on port 9294
arena.run_server(port=9294)
```

Based on the port provided, the DALEX server will be running on `https://arena.`
`drwhy.ai/?data=http://127.0.0.1:9294/`. Initially, you will get a blank
dashboard, but you can easily drag and drop the visuals from the right-hand side panel
to make your own custom dashboard, as shown in the following screenshot:

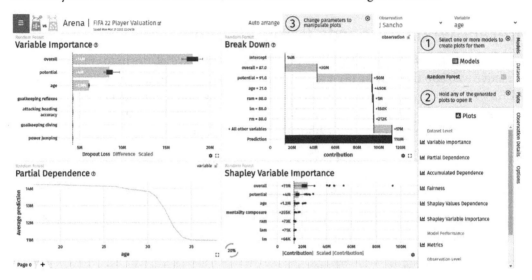

Figure 9.8 – An interactive Arena dashboard created using DALEX

Also, you can load an existing dashboard from a configuration JSON or export a build
dashboard as a configuration JSON file. Try recreating the dashboard, as shown in
Figure 9.8, using the configuration JSON file provided in the code repository at
`https://raw.githubusercontent.com/PacktPublishing/Applied-`
`Machine-Learning-Explainability-Techniques/main/Chapter09/`
`dalex_sessions/session-1647894542387.json`.

Overall, I have found DALEX to be a very interesting and powerful XAI framework. There
are many more examples available at `https://github.com/ModelOriented/`
`DALEX` and `https://github.com/ModelOriented/DrWhy/blob/master/`
`README.md`. Please do explore all of them. However, DALEX seems to be restricted to
structured data. I think as a future scope, making DALEX easily applicable with image
and text data would increase its adoption across the AI research community. In the next
section, we will explore Explainerdashboard, which is another interesting XAI framework.

Explainerdashboard

The AI research community has always considered interactive visualization to be an important approach for interpreting ML model predictions. In this section, we will cover **Explainerdashboard**, which is an interesting Python framework that can spin up a comprehensive interactive dashboard covering various aspects of model explainability with just minimal lines of code. Although this framework supports only scikit-learn-compatible models (including XGBoost, CatBoost, and LightGBM), it can provide model-agnostic global and local explainability. Currently, it supports SHAP-based feature importance and interactions, PDPs, model performance analysis, what-if model analysis, and even decision-tree-based breakdown analysis plots.

The framework allows customization of the dashboard, but I think the default version includes all supported aspects of model explainability. The generated web-app-based dashboards can be exported as static web pages directly from a live dashboard. Otherwise, the dashboards can be programmatically deployed as a web app through an automated **Continuous Integration (CI)**/**Continuous Deployment (CD)** deployment process. I recommend that you go through the official documentation of the framework (`https://explainerdashboard.readthedocs.io/en/latest/`) and the GitHub project repository (`https://github.com/oegedijk/explainerdashboard`) before we get started with the walk-through tutorial example next.

Setting up Explainerdashboard

The complete tutorial notebook is provided in the code repository for this chapter at `https://github.com/PacktPublishing/Applied-Machine-Learning-Explainability-Techniques/blob/main/Chapter09/Explainer_dashboard_example.ipynb`. However, in this section, I will provide a complete walk-through of the tutorial. The same *FIFA Club Position Prediction dataset* (`https://www.kaggle.com/datasets/adityabhattacharya/fifa-club-position-prediction-dataset`) will be used for this tutorial, too. But instead of using the dataset to predict the valuation of football players, here, I will use this dataset to predict the league position of the football club for the next season based on the skills and ability of the football players playing for the club.

The real-world task of predicting a club league position for a future season is more complex, and there are several other variables that need to be included to get an accurate prediction. However, this prediction problem is solely based on the quality of the players playing for the club.

To get started with the tutorial, you will need to install all of the required dependencies to run the notebook. If you have executed all the previous tutorial examples, then most of the Python modules should be installed, except for **Explainerdashboard**. You can install explainerdashboard using the pip installer:

```
!pip install explainerdashboard
```

The Explainerdashboard framework does have a dependency on the graphviz module, which makes it slightly tedious to install depending on your system. At the time of writing, I have discovered that version 0.18 works best with Explainerdashboard. This can be installed using the pip installer:

```
!pip install graphviz==0.18
```

Graphviz is an open source graph visualization software that is needed for the decision tree breakdown plot used in Explainerdashboard. In spite of the pip installer, you might also need to install the graphviz binaries depending on the operating system that you are using. Please visit https://graphviz.org/ to find out more. Additionally, if you are facing any friction during the setup of this module, take a look at the installation instructions provided at https://pypi.org/project/graphviz/.

We will consider this ML problem to be a regression problem. Therefore, similar to the DALEX example, we will need to perform the same data preprocessing, feature engineering, model training, and evaluation steps. I recommend that you follow the steps provided in the notebook at https://github.com/PacktPublishing/Applied-Machine-Learning-Explainability-Techniques/blob/main/Chapter09/Explainer_dashboard_example.ipynb. This contains the necessary details to get the trained model. We will use this trained model as a black box and use Explainerdashboard to explain it in the next section.

Model explainability with Explainerdashboard

After the installation of the Explainerdashboard Python module is successful, you can import it to verify the installation:

```
import explainerdashboard
```

For this example, we will use the `RegressionExplainer` and `ExplainerDashboard` submodules. So, we will load the specific submodules:

```
from explainerdashboard import RegressionExplainer,
ExplainerDashboard
```

Next, using just two lines of code, we can spin up the `ExplainerDashboard` submodule for this example:

```
explainer = RegressionExplainer(model _ skl, x _ valid, y _ valid)
ExplainerDashboard(explainer).run()
```

Once this step is running successfully, the dashboard should be running in `localhost` with port `8050` as the default port. So, you can visit `http://localhost:8050/` in the browser to view your explainer dashboard.

The following lists the different explainability methods provided by Explainerdashboards:

- **Feature importance**: Similar to other XAI frameworks, feature importance is an important method for gaining an understanding of the overall contribution of each attribute used for prediction. This framework uses SHAP values, permutation importance, and PDPs to analyze the contribution of each feature for the model prediction:

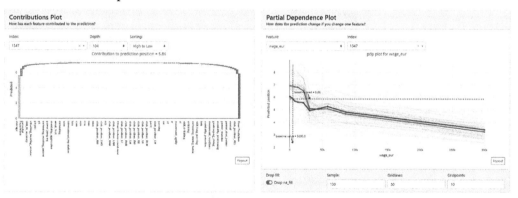

Figure 9.9 – Contribution plots and PDPs from Explainerdashboard

- **Model performance**: Similar to DALEX, Explainerdashboard also allows you to analyze the model performance. For classification models, it uses metrics such as precision plots, confusion matrices, ROC-AUC plots, PR AUC plots, and more. For regression models, we will see plots such as residual plots, goodness-of-fit plots, and more:

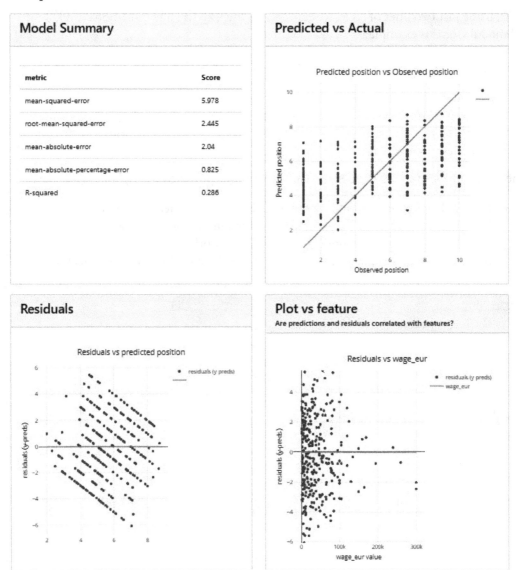

Figure 9.10 – Model performance analysis plots for regression models in Explainerdashboard

- **Prediction-level analysis**: Explainerdashboard provides interesting and interactive plots for getting local explanations. This is quite similar to other Python frameworks. It is very important to have for analyzing prediction-level outcomes.

- **What-if analysis**: Another interesting option that Explainerdashboard provides is the what-if analysis feature. We can use this feature to vary the feature values and observe how the overall prediction gets changed. I find what-if analysis to be very useful for providing prescriptive insights:

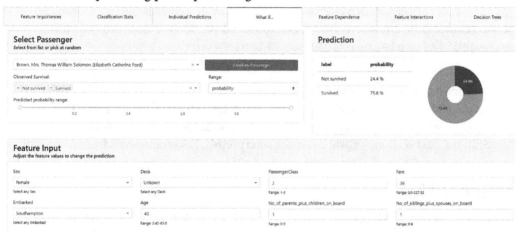

Figure 9.11 – What-if model analysis using Explainerdashboard

- **Feature dependence and interactions**: Analyzing the dependency and interactions between different features is another interesting explainability method provided in Explainerdashboard. Mostly, it uses SHAP methods for analyzing feature dependence and interactions.

- **Decision tree surrogate explainers**: Explainerdashboard uses decision trees as surrogate explainers. Additionally, it uses the decision tree breakdown plot for model explainability:

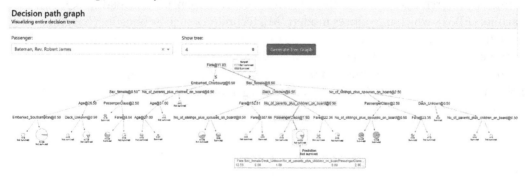

Figure 9.12 – Decision tree surrogate explainers in Explainerdashboard

To stop running the dashboards on your local system, you can simply interrupt the notebook cell.

Explainerdashboard offers you many customization options as well. To customize your own dashboard from the given template, it is recommended that you refer to `https://github.com/oegedijk/explainerdashboard#customizing-your-dashboard`. You can also build multiple dashboards and compile all the dashboards as an explainer hub: `https://github.com/oegedijk/explainerdashboard#explainerhub`. To deploy dashboards into a live web app that is accessible from anywhere, I would recommend you look at `https://github.com/oegedijk/explainerdashboard#deployment`.

In comparison to DALEX, I would say that Explainerdashboard is slightly behind as it is only restricted to scikit-learn-compatible models. This means that with complex deep learning models built on unstructured data such as images and text, you can't use this framework. However, I found it easy to use and very useful for ML models built on tabular datasets. In the next section, we will cover the InterpretML XAI framework from Microsoft.

InterpretML

InterpretML (`https://interpret.ml/`) is an XAI toolkit from Microsoft. It aims to provide a comprehensive understanding of ML models for the purpose of model debugging, outcome explainability, and regulatory audits of ML models. With this Python module, we can either train *interpretable glassbox models* or *explain black-box models*.

In *Chapter 1*, *Foundational Concepts of Explainability Techniques*, we discovered that some models such as decision trees, linear models, or rule-fit algorithms are inherently explainable. However, these models are not efficient for complex datasets. Usually, these models are termed glass-box models as opposed to black-box models, as they are extremely transparent.

Microsoft Research developed another algorithm called **Explainable Boosting Machine** (**EBM**), which introduces modern ML techniques such as boosting, bagging, and automatic interaction detection into classical algorithms such as **Generalized Additive Models** (**GAMs**). Researchers have also found that EBMs are accurate as random forests and gradient-boosted trees, but unlike such black-box models, EBMs are explainable and transparent. Therefore, EBMs are glass-box models that are built into the InterpretML framework.

In comparison to DALEX and Explainerdashboard, InterpretML is slightly behind in terms of both usability and adoption. However, since this framework as a great potential to evolve further, it is important to discuss this framework. Before discussing the code tutorial, let us discuss about the explainability techniques that are supported by this framework.

Supported explanation methods

At the time of writing, the following table illustrates the supported explanation methods in InterpretML, as mentioned in the GitHub project source at `https://github.com/interpretml/interpret#supported-techniques`:

Explanation Algorithms	Explainability Type
Explainable Boosting	Glassbox Model
Decision Trees	Glassbox Model
Rule-Fit	Glassbox Model
Linear/Logistic Regression	Glassbox Model
SHAP Kernel Explainer	Black-box Model
LIME	Black-box Model
Morris Sensitivity	Black-box Model
Partial Dependence	Black-box Model

Figure 9.13 – Explanation methods supported in InterpretML

I recommend that you keep an eye on the project documentation, as I am quite certain the supported explanation methods for this framework will increase for InterpretML. Next, let's explore how to use this framework in practice.

Setting up InterpretML

In this section, I will walk you through the tutorial example of InterpretML that is provided in the code repository at `https://github.com/PacktPublishing/Applied-Machine-Learning-Explainability-Techniques/blob/main/Chapter09/InterpretML_example.ipynb`. In the tutorial, we used InterpretML to explain an ML model trained for hepatitis detection, which is a classification problem.

To begin the problem, you need to have the InterpretML Python module installed. You can use the pip installer for this:

```
pip install interpret
```

Although the framework is supported by Windows, Mac, and Linux, it does require you to have a Python version that is higher than 3.6. You can validate whether the installation is successful by importing the module:

```
import interpret as iml
```

Next, let's discuss the dataset that is used in this tutorial.

Discussions about the dataset

The hepatitis detection dataset is taken from the UCI Machine Learning repository at `https://archive.ics.uci.edu/ml/datasets/hepatitis`. It has 155 records and 20 features of different types for the detection of the hepatitis disease. Therefore, this dataset is used for solving binary classification problems. For your convenience, I have added this dataset to the code repository at `https://github.com/PacktPublishing/Applied-Machine-Learning-Explainability-Techniques/tree/main/Chapter09/datasets/Hepatitis_Data`.

> **Hepatitis Dataset Citation**
> *G.Gong (Carnegie-Mellon University) via Bojan Cestnik, Jozef Stefan Institute* (`https://archive.ics.uci.edu/ml/datasets/hepatitis`)

More details about the dataset and initial exploration results are included in the tutorial notebook. However, on a very high level, *the dataset is imbalanced*, it has *missing values*, and it has both *categorical* and *continuous* variables. Therefore, it needs necessary transformation before the model can be built. All these necessary steps are included in the tutorial notebook, but please feel free to explore additional methods for building a better ML model.

Training the model

For this example, after dividing the entire data into a training set and a test set, I have trained a random forest classifier with minimum hyperparameter tuning:

```
x _ train, x _ test, y _ train, y _ test = train _ test _ split(
    encoded, label, test _ size=0.3, random _ state=123)
model = RandomForestClassifier(
  n _ estimators=500, min _ samples _ split = 3,
  random _ state=123).fit(x _ train, y _ train)
```

Note that sufficient hyperparameter tuning is not done for this model, as we are more interested in the model explainability part with InterpretML rather than learning how to build an efficient ML model. However, I encourage you to explore hyperparameters tuning further to get a better model.

On evaluating the model on the test data, we have received an accuracy of 85% and an **Area Under the ROC Curve (AUC)** score of 70%. The AUC score is much lower than the accuracy as the dataset used is imbalanced. This indicates that a metric such as accuracy can be misleading. Therefore, it is better to consider metrics such as the AUC score, F1 score, and confusion matrix instead of accuracy for model evaluation.

Next, we will use InterpretML for model explainability.

Explainability with InterpretML

As mentioned earlier, with InterpretML. you can either use interpretable glass-box models as surrogate explainers or explore certain model-agnostic methods to explain black-box models. With both approaches, you can get an interactive dashboard for analyzing the various aspects of explainability. First, I will cover the model explainability using glass-box models in InterpretML.

Explaining with glass-box models using InterpretML

InterpretML supports interpretable glass-box models such as the **Explainable Boosting Machine (EBM)**, **Decision Tree**, and **Rule-Fit** algorithms. These algorithms are applied as surrogate explainers for providing post hoc model explainability. First, let's try out the EBM algorithm.

EBM

To explain a model with EBM, we need to load the required submodule in Python:

```
from interpret.glassbox import ExplainableBoostingClassifier
```

Once the EBM submodule has been successfully imported, we just need to create a trained surrogate explainer object:

```
ebm = ExplainableBoostingClassifier(feature_types=feature_types)
ebm.fit(x_train, y_train)
```

The ebm variable is the EBM explainer object. We can use this variable to get global or local explanations. The framework only supports feature importance-based global and local explainability but creates an interactive plot for further analysis:

```
# Showing Global Explanations
ebm_global = ebm.explain_global()
iml.show(ebm_global)
# Local explanation using EBM
ebm_local = ebm.explain_local(x_test[5:6], y_test[5:6],
                              name = 'Local Explanation')
iml.show(ebm_local)
```

Figure 9.14 illustrates the global feature importance plot and variation of the *Age* feature with the overall data distribution obtained using InterpretML:

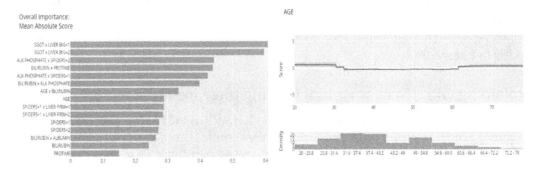

Figure 9.14 – Global explanation plots using InterpretML

Feature importance for the local explanation, which has been done at the prediction level of the individual data instance, is shown in *Figure 9.15*:

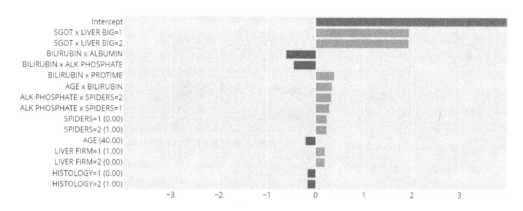

Figure 9.15 – Local explanation using InterpretML

Next, we will explore rule-based algorithms in InterpretML as *surrogate explainers*, as discussed in *Chapter 2*, *Model Explainability Methods*.

Decision rule list

Similar to EBM, another popular glass-box surrogate explainer that is available in InterpretML is the decision rule list. This is similar to the rule-fit algorithm, which can learn specific rules from the dataset to explain the logical working of the model. We can apply this method using InterpretML in the following way:

```
from interpret.glassbox import DecisionListClassifier
dlc = DecisionListClassifier(feature_types=feature_types)
dlc.fit(x_train, y_train)

# Showing Global Explanations
dlc_global = dlc.explain_global()
iml.show(dlc_global)
```

With this method, the framework displays the learned rules, as shown in the following screenshot:

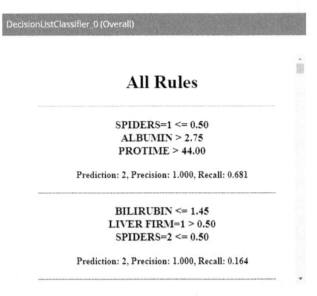

Figure 9.16 – Decision rule list using InterpretML

As we can see in *Figure 9.16*, it generates a list of learned rules. Next, we will explore the decision tree-based surrogate explainer in InterpretML.

Decision tree

Similar to a decision rule list, we can also fit the decision tree algorithm as a surrogate explainer using InterpretML for model explainability. The API syntax is also quite similar for applying a decision tree classifier:

```
from interpret.glassbox import ClassificationTree
dtc = ClassificationTree(feature_types=feature_types)
dtc.fit(x_train, y_train)

# Showing Global Explanations
dtc_global = dtc.explain_global()
iml.show(dtc_global)
```

This produces a decision tree breakdown plot, as shown in the following screenshot, for the model explanation:

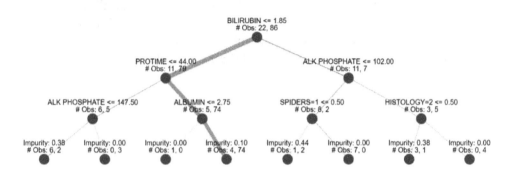

Figure 9.17 – A decision tree-based surrogate explainer in InterpretML

Now all of these individual components can also be clubbed together into one single dashboard using the following line of code:

```
iml.show([ebm_global, ebm_local, dlc_global, dtc_global])
```

Figure 9.18 illustrates the consolidated interactive dashboard using InterpretML:

Figure 9.18 – The InterpretML dashboard consolidating all individual plots

In the next section, we will cover the various methods that are available in InterpretML for providing a model-agnostic explanation of the black-box model.

Explaining black-box models using InterpretML

In this section, we will cover the four different methods supported in InterpretML for explaining black-box models. We will only cover the code part as the visualizations of feature importance and feature variation are very similar to the glass-box models. I do recommend looking at the tutorial notebook for interacting with the generated plots to gain more insights. The methods supported are LIME, Kernel SHAP, Morris Sensitivity, and Partial Dependence:

```
from interpret.blackbox import LimeTabular, ShapKernel,
MorrisSensitivity, PartialDependence
```

First, we will explore the LIME tabular method:

```
#The InterpretML Blackbox explainers need a predict function,
and optionally a dataset
lime = LimeTabular(predict_fn=model.predict_proba, data=x_
train.astype('float').values, random_state=123)
#Select the instances to explain, optionally pass in labels if
you have them
lime_local = lime.explain_local(
    x_test[:5].astype('float').values,
    y_test[:5], name='LIME')
```

Next, InterpretML provides the SHAP Kernel method for a model-agnostic SHAP-based explanation:

```
# SHAP explanation
background_val = np.median(
    x_train.astype('float').values, axis=0).reshape(1, -1)
shap = ShapKernel(predict_fn=model.predict_proba,
                  data=background_val,
                  feature_names=list(x_train.columns))
shap_local = shap.explain_local(
    x_test[:5].astype('float').values,
    y_test[:5], name='SHAP')
```

Another model-agnostic global explanation method that is supported is **Morris Sensitivity**, which is used to obtain the overall sensitivity of the features:

```
# Morris Sensitivity
sensitivity = MorrisSensitivity(
   predict_fn=model.predict_proba,
   data=x_train.astype('float').values,
   feature_names=list(x_train.columns),
   feature_types=feature_types)
sensitivity_global = sensitivity.explain_global(name="Global
Sensitivity")
```

InterpretML also supports PDPs for analyzing feature dependence:

```
# Partial Dependence
pdp = PartialDependence(
   predict_fn=model.predict_proba,
   data=x_train.astype('float').values,
   feature_names=list(x_train.columns),
   feature_types=feature_types)
pdp_global = pdp.explain_global(name='Partial Dependence')
```

Finally, everything can be consolidated into a single dashboard using a single line of code:

```
iml.show([lime_local, shap_local, sensitivity_global,
          pdp_global])
```

This will create a similar interactive dashboard, as shown in *Figure 9.18.*

With the various surrogate explainers and interactive dashboards, this framework does have a lot of potential, even though there are quite a few limitations. It is restricted to tabular datasets, it is not compatible with model frameworks such as PyTorch, TensorFlow, and H20, and I think the model explanation methods are also limited. Improving these limitations can definitely increase the adoption of this framework.

Next, we will cover another popular XAI framework – ALIBI for model explanation.

ALIBI

ALIBI is another popular XAI framework that supports both local and global explanations for classification and regression models. In *Chapter 2, Model Explainability Methods*, we did explore this framework for getting counterfactual examples, but ALIBI does include other model explainability methods too, which we will explore in this section. Primarily, ALIBI is popular for the following list of model explanation methods:

- **Anchor explanations**: An anchor explanation is defined as a rule that sufficiently revolves or anchors around the local prediction. This means that if the anchor value is present in the data instance, the model prediction is almost always the same, irrespective of changes to other feature values.

- **Counterfactual Explanations (CFEs)**: We have seen counterfactuals in *Chapter 2, Model Explainability Methods*. CFEs indicate which feature values should change, and by how much, to produce a different outcome.

- **Contrastive Explanation Methods (CEMs)**: CEMs are used with classification models for local explanations in terms of **Pertinent Positives (PPs)**, meaning features that should be minimally and sufficiently present to justify a given classification, and **Pertinent Negatives (PNs)**, meaning features that minimally and necessarily absent to justify the classification.

- **Accumulated Local Effects (ALE)**: ALE plots illustrate how attributes influence the overall prediction of an ML model. ALE plots are often considered to be unbiased and a faster alternative to PDPs, as covered in *Chapter 2, Model Explainability Methods*.

To get a detailed summary of the supported methods for model explanation, please take a look at `https://github.com/SeldonIO/alibi#supported-methods`. Please explore the official documentation of this framework to learn more about it: `https://docs.seldon.io/projects/alibi/en/latest/examples/overview.html`.

Now, let me walk you through the code tutorial provided for ALIBI.

Setting up ALIBI

The complete code tutorial is provided in the project repository for this chapter at `https://github.com/PacktPublishing/Applied-Machine-Learning-Explainability-Techniques/blob/main/Chapter09/ALIBI_example.ipynb`. If you have followed the tutorials for *Counterfactual explanations* from *Chapter 2, Model Explainability Methods*, you should have ALIBI installed already.

You can import the submodules that we are going to use for this example from ALIBI:

```
import alibi
from alibi.explainers import AnchorTabular, CEM,
CounterfactualProto, ale
```

Next, let's discuss the dataset for this tutorial.

Discussion about the dataset

For this example, we will use the *Occupancy Detection* dataset from the *UCI Machine Learning* repository at `https://archive.ics.uci.edu/ml/datasets/Occupancy+Detection+#`. This dataset is used for detecting whether a room is occupied or not from the different sensor values that are provided. Hence, this is a classification problem that can be solved by fitting ML classifiers on the given dataset. The detailed data inspection, preprocessing, and transformation steps are included in the tutorial notebook. On a very high level, the dataset is slightly imbalanced and, mostly, contains numerical features with no missing values.

> **Occupancy Detection dataset citation**
>
> L.M. Candanedo, V. Feldheim (2016) - Accurate occupancy detection of an office room from light, temperature, humidity and CO2 measurements using statistical learning models. (`https://archive.ics.uci.edu/ml/datasets/Occupancy+Detection+#`)

In this tutorial, I have demonstrated how to use a pipeline approach with scikit-learn for training ML models. This is a very neat way of building ML models, and it is especially useful when working on industrial problems that need to be deployed to the production system. To learn more about this approach, take a look at the official scikit-learn pipeline documentation at `https://scikit-learn.org/stable/modules/generated/sklearn.pipeline.Pipeline.html`.

Next, let's discuss the model that will be used for extracting explanations.

Training the model

For this example, I have used a random forest classifier to train a model with minimal hyperparameter tuning. You can explore other ML classifiers too, as the choice of the algorithm doesn't matter. Our goal is to explore ALIBI for model explainability, which I will cover in the next section.

Model explainability with ALIBI

Now, let's use the various explanation methods discussed earlier for the trained model, which we can consider a black box.

Using anchor explanations

In order to get the anchor points, first, we need to create an anchor explanation object:

```
explainer = AnchorTabular(
    predict_fn,
    feature_names=list(df_train.columns),
    seed=123)
```

Next, we need to fit the explainer object on the training data:

```
explainer.fit(df_train.values, disc_perc=[25, 50, 75])
```

We need to learn an anchor value for both the occupied class and the unoccupied class. This process involves providing a data instance belonging to each of these classes as input for estimating the anchor points. This can be done by using the following lines of code:

```
class_names = ['not_occupied', 'occupied']
print('Prediction: ',
      class_names[explainer.predictor(
          df_test.values[5].reshape(1, -1))[0]])
explanation = explainer.explain(df_test.values[5],
                                threshold=0.8)
print('Anchor: %s' % (' AND '.join(explanation.anchor)))
print('Prediction: ',
      class_names[explainer.predictor(
          df_test.values[100].reshape(1, -1))[0]])

explanation = explainer.explain(df_test.values[100],
                                threshold=0.8)
print('Anchor: %s' % (' AND '.join(explanation.anchor)))
```

In this example, the anchor point for the occupied class is obtained when the light intensity value is greater than 256.7 and the CO2 value is greater than 638.8. In comparison, for the unoccupied class, it is obtained when the CO2 value is greater than 439. Essentially, this is telling us that if the sensor values measuring light intensity are greater than 256.7 and the CO2 levels are greater than 638.8, the model predicts that the room is occupied.

The pattern learned by the model is actually appropriate, as whenever a room is occupied, it is more likely that the lights are turned on, and with more occupants, CO2 levels are also likely to increase. The anchor points for the unoccupied class are not very appropriate, intuitive, or interpretable, but this indicates that, usually, CO2 levels are lower when the room is not occupied. Typically, we get to learn about some threshold values of certain impact features that the model relies on for predicting the outcome.

Using CEM

With CEM, the main idea is to learn PPs or conditions that should be present to justify the occurrence of one class and PNs, which should be absent to indicate the occurrence of one class. This is used for model-agnostic local explainability. You can find out more about this method from this research literature at https://arxiv.org/pdf/1802.07623.pdf.

To apply this in Python, we need to create a CEM object with the required hyper-parameters and fit the train values to learn the PP and PN values:

```
cem = CEM(predict_fn, mode, shape, kappa=kappa,
          beta=beta, feature_range=feature_range,
          max_iterations=max_iterations, c_init=c_init,
          c_steps=c_steps,
          learning_rate_init=lr_init, clip=clip)
cem.fit(df_train.values, no_info_type='median')
explanation = cem.explain(X, verbose=False)
```

In our example, the PP and PN values that have been learned show by how much the value should be increased or decreased to meet the minimum criteria for the correct outcome. Surprisingly, no PN value was obtained for our example. This indicates that all the features are important for the model. The absence of any feature or any particular value range does not help the model predict the outcome. Usually, for higher-dimensional data, the PN values would be important to analyze.

Using CFEs

In *Chapter 2, Model Explainability Methods*, we looked at tutorial examples of how CFEs can be applied with ALIBI. We will follow a similar approach for this example, too. However, ALIBI does allow different algorithms to generate CFEs, which I highly recommend you explore: `https://docs.seldon.io/projects/alibi/en/latest/methods/CF.html`. In this chapter, we will stick to the prototype-based method covered in the CFE tutorial of *Chapter 2, Model Explainability Methods*:

```
cfe = CounterfactualProto(predict_fn,
                          shape,
                          use_kdtree=True,
                          theta=10.,
                          max_iterations=1000,
                          c_init=1.,
                          c_steps=10
                          )
cfe.fit(df_train.values, d_type='abdm',
        disc_perc=[25, 50, 75])
explanation = cfe.explain(X)
```

Once the explanation object is ready, we can actually compare the difference between CFEs and the original data instance to understand the change in the feature values required to flip the outcome. However, using this method to get the correct CFE can be slightly challenging as there are many hyperparameters that require the right tuning; therefore, the method can be challenging and tedious. Next, let's discuss how to use ALE plots for model explainability.

Using ALE plots

Similar to PDPs, as covered in *Chapter 2, Model Explainability Methods*, ALE plots can be used to find the relationship of the individual features with respect to the target class. Let's see how to apply this in Python:

```
proba_ale = ale.ALE(predict_fn, feature_names=numeric,
                    target_names=class_names)
proba_explain = proba_ale.explain(df_test.values)
ale.plot_ale(proba_explain, n_cols=3,
             fig_kw={'figwidth': 12, 'figheight': 8})
```

This will create the following ALE plots:

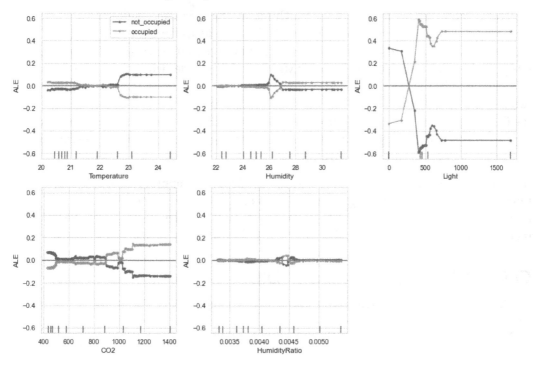

Figure 9.19 – ALE plots using ALIBI

In *Figure 9.19*, we can see that the variance in feature values for the occupied and not occupied target classes is at the maximum level for the feature light, followed by CO2 and temperature, and at the lowest level for HumidityRatio. This gives us an indication of how the model prediction changes depending on the variation of the feature values.

Overall, I feel that ALIBI is an interesting XAI framework that works with tabular and unstructured data such as text and images and does have a wide variety of techniques for the explainability of ML models. The only limitation I have found is that some of the methods are not very simplified, so they require a good amount of hyperparameter tuning to get reliable explanations. Please explore https://github.com/SeldonIO/alibi/tree/master/doc/source/examples for other examples provided for ALIBI to get more practical expertise. In the next section, we will discuss DiCE as an XAI Python framework.

DiCE

Diverse Counterfactual Explanations (**DiCE**) is another popular XAI framework that we briefly covered in *Chapter 2*, *Model Explainability Methods*, for the *CFE tutorial*. Interestingly, DiCE is also one of the key XAI frameworks from Microsoft Research, but it is yet to be integrated with the InterpretML module (I wonder why!). I find the entire idea of CFE to be very close to the ideal human-friendly explanation that gives actionable recommendations. This blog from Microsoft discusses the motivation and idea behind the DiCE framework: `https://www.microsoft.com/en-us/research/blog/open-source-library-provides-explanation-for-machine-learning-through-diverse-counterfactuals/`.

In comparison to ALIBI CFE, I found DiCE to produce more appropriate CFEs with minimal hyperparameter tuning. That's why I feel it's important to mention DiCE, as it is primarily designed for example-based explanations. Next, let's discuss the CFE methods that are supported in DiCE.

CFE methods supported in DiCE

DiCE can generate CFEs based on the following methods:

- Model-agnostic methods:

  - KD-Tree

  - Genetic algorithm

  - Randomized sampling

- Gradient-based methods (model-specific methods):

  - Loss-based method for deep learning models

  - **Variational Auto-Encoder** (**VAE**)-based methods

To learn more about all these methods, I request that you explore the official documentation of DiCE (`https://github.com/interpretml/DiCE`), which contains the necessary research literature for each method. Now, let's use DiCE for model explainability.

Model explainability with DiCE

The complete tutorial example is provided at `https://github.com/ PacktPublishing/Applied-Machine-Learning-Explainability- Techniques/blob/main/Chapter09/DiCE_example.ipynb`. For this example, I have used the same Occupancy Detection dataset that we used for the ALIBI tutorial. Since the same data preprocessing, transformation, model training, and evaluation steps have been used, we will directly proceed with the model explainability part with DALEX. The notebook tutorial contains all the necessary steps, so I recommend that you go through the notebook first.

We will use the DiCE framework in the same way as we have done for the CFE tutorial from *Chapter 2, Model Explainability Methods*.

So, first, we need to define a DiCE data object:

```
data _ object = dice _ ml.Data(
   dataframe = df _ train[numeric + [target _ variable]],
   continuous _ features = numeric,
   outcome _ name = target _ variable
)
```

Next, we need to create a DiCE model object:

```
model _ object = dice _ ml.Model(model=model,backend='sklearn')
```

Following this, we need to pass the data object and the model object for the DiCE explanation object:

```
explainer = dice _ ml.Dice(data _ object, model _ object,
                           method = 'random')
```

Next, we can take a query data instance and generate CFEs using the DiCE explainer object:

```
test _ query = df _ test[400:401][numeric]

cfe = explainer.generate _ counterfactuals(
     test _ query,
     total _ CFs=4,
     desired _ range=None,
     desired _ class="opposite",
```

```
    features _ to _ vary= numeric,
    permitted _ range = { 'CO2' : [400, 1000]}, # Adding a
constraint for CO2 feature
    random _ seed = 123,
    verbose=True)


cfe.visualize _ as _ dataframe(show _ only _ changes=True)
```

This will produce a CFE DataFrame that shows the feature values that need to be changed to flip the model predicted outcome. The outcome of this approach is illustrated in the following screenshot:

```
Diverse Counterfactuals found! total time taken: 00 min 01 sec
Query instance (original outcome : 0)
```

	Temperature	Humidity	Light	CO2	HumidityRatio	Occupancy
0	20.89	23.1	0.0	499.666667	0.003523	0

```
Diverse Counterfactual set (new outcome: 1.0)
```

	Temperature	Humidity	Light	CO2	HumidityRatio	Occupancy
0	-	24.000000000000014	763.5	696.3	-	1.0
1	-	24.000000000000014	1083.9	500.0666667000001	-	1.0
2	-	24.000000000000014	1046.8	500.0666667000001	0.005247647	1.0
3	-	24.000000000000014	908.6	500.0666667000001	0.004907888	1.0

Figure 9.20 – CFE generated using the DiCE framework, which is displayed as a DataFrame

Interestingly, CFEs don't only provide actionable insights from the data. However, they can also be used to generate local and global feature importance. The features that can be easily varied to alter the model prediction are considered to be more important by this approach of feature importance. Let's try applying the local feature importance using the DiCE method:

```
local _ importance = explainer.local _ feature _ importance(test _
query)
print(local _ importance.local _ importance)
plt.figure(figsize=(10,5))
plt.bar(range(len(local _ importance.local _ importance[0])),
        list(local _ importance.local _ importance[0].values())/(np.
sum(list(local _ importance.local _ importance[0].values()))),
```

```
          tick _ label=list(local _ importance.local _ importance[0].
keys()),
          color = list('byrgmc')
       )
plt.show()
```

This produces the following local feature importance plot for the test query selected:

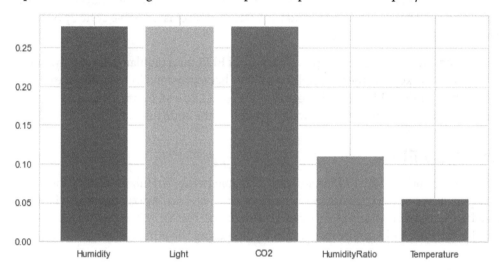

Figure 9.21 – Local feature importance using DiCE

Figure 9.21 shows that for the test query data, the humidity, light, and CO2 features are the most important for model prediction. This indicates that most CFEs would suggest changing the feature values of one of these features to alter the model prediction.

Overall, DiCE is a very promising framework for robust CFEs. I recommend you explore the different algorithms to generate CFEs such as KD-Trees, random sampling, and genetic algorithms. DiCE examples can sometimes be very random. My recommendation is to always use a random seed to control the randomness, clearly define the actionable and non-actionable features, and set the boundary conditions of the actionable features to generate CFEs that are meaningful and practically feasible. Otherwise, the generated CFEs can be very random and practically not feasible and, therefore, less impactful to use.

For other examples of the DiCE framework for multiclass classification or regression problems, please explore https://github.com/interpretml/DiCE/tree/master/docs/source/notebooks. Next, let's cover ELI5, which is one of the initial XAI frameworks that has been developed to produce simplistic explanations of ML models.

ELI5

ELI5, or *Explain Like I'm Five*, is a Python XAI library for debugging, inspecting, and explaining ML classifiers. It was one of the initial XAI frameworks developed to explain black-box models in the most simplified format. It supports a wide range of ML modeling frameworks such as scikit-learn compatible models, Keras, and more. It also has integrated LIME explainers and can work with tabular datasets along with unstructured data such as text and images. The library documentation is provided at `https://eli5.readthedocs.io/en/latest/`, and the GitHub project is available at `https://github.com/eli5-org/eli5`.

In this section, we will cover the application part of ELI5 for a tabular dataset only, but please feel free to explore other examples that have been provided in the tutorial examples of ELI5 at `https://eli5.readthedocs.io/en/latest/tutorials/index.html`. Next, let's get started with the walk-through of the code tutorial.

Setting up ELI5

The complete tutorial of the ELI5 example is available in the GitHub repository for this chapter: `https://github.com/PacktPublishing/Applied-Machine-Learning-Explainability-Techniques/blob/main/Chapter09/ELI5_example.ipynb`. ELI5 can be installed in Python using the pip installer:

```
pip install eli5
```

If the installation process is successful, you can verify it by importing the module in the Jupyter notebook:

```
import eli5
```

For this example, we will use the same hepatitis detection dataset from the UCI Machine Learning repository (`https://archive.ics.uci.edu/ml/datasets/hepatitis`), which we used for the InterpretML example. Also, we have used a random forest classification model with minimum hyperparameter tuning that will be used as our black-box model. So, we will skip the discussion about the dataset and model part and proceed to the model explainability part using ELI5.

Model explainability using ELI5

Applying ELI5 in Python is very easy and can be done with a few lines of code:

```
eli5.show_weights(model, vec = DictVectorizer(),
                  feature_names = list(encoded.columns))
```

This will produce the following feature weight tabular visualization that can be used to analyze the global feature importance:

Weight	Feature
0.1273 ± 0.2324	BILIRUBIN
0.0926 ± 0.2142	ALBUMIN
0.0857 ± 0.1815	ALK PHOSPHATE
0.0801 ± 0.1705	PROTIME
0.0653 ± 0.1381	SGOT
0.0629 ± 0.1471	AGE
0.0487 ± 0.1553	SPIDERS=1
0.0477 ± 0.1555	SPIDERS=2
0.0351 ± 0.1186	LIVER FIRM=1
0.0343 ± 0.1298	LIVER FIRM=2
0.0262 ± 0.1058	HISTOLOGY=2
0.0248 ± 0.1046	SPLEEN PALPABLE=1
0.0245 ± 0.1157	ASCITES=1
0.0233 ± 0.1078	ASCITES=2
0.0225 ± 0.0909	MALAISE=1
0.0210 ± 0.0880	SPLEEN PALPABLE=2
0.0205 ± 0.0844	HISTOLOGY=1
0.0194 ± 0.0882	MALAISE=2
0.0161 ± 0.0676	ANOREXIA=1
0.0144 ± 0.0636	STEROID=1
... 12 more ...	

Figure 9.22 – Feature weights obtained using ELI5

Figure 9.22 indicates that the feature, BILIRUBIN, has the maximum weight and, hence, has the maximum contribution for influencing the model outcome. The +/- values shown beside the weight values can be considered to be confidence intervals. This method can be considered a very simple way to provide insights into the black-box model. ELI5 calculates the feature weights using tree models. Every node of the tree gives an output score that is used to estimate the total contribution of a feature. The total contribution on the decision path is how much the score changes from parent to child. The total weights of all the features sum up the total probability of the model for predicting a particular class.

We can use this method for providing local explainability and for an inference data instance:

```
no_missing = lambda feature_name, feature_value: not
np.isnan(feature_value) # filter missing values
eli5.show_prediction(model,
                x_test.iloc[1:2].astype('float'),
                feature_names = list(encoded.columns),
                show_feature_values=True,
                feature_filter=no_missing,
                target_names = {1:'Die', 2:'Live'},
                top = 10,
```

```
show = ['feature _ importances',
            'targets', 'decision _ tree',
            'description'])
```

This will produce the following tabular visualization for analyzing the feature contributions of the inference data:

y=Live (probability **0.984**) top features

Contribution[?]	Feature	Value
+0.801	<BIAS>	1.000
+0.030	SPIDERS=2	1.000
+0.027	SPIDERS=1	0.000
+0.021	BILIRUBIN	0.700
+0.017	ALBUMIN	4.200
+0.010	HISTOLOGY=2	0.000
+0.009	MALAISE=1	0.000
+0.008	MALAISE=2	1.000
+0.008	FATIGUE=1	0.000
… 19 more positive …		
… 4 more negative …		
-0.016	SGOT	224.000

Figure 9.23 – Feature contributions using ELI5 for local explainability

In *Figure 9.23*, we can see the feature contributions using ELI5 for the local data used for prediction. There is a `<BIAS>` term that is added to the table. This is considered the expected average score output by the model, which depends on the distribution of the training data. To find out more, take a look at this Stack Overflow post: `https://stackoverflow.com/questions/49402701/eli5-explaining-prediction-xgboost-model`.

Even though ELI5 is easy to use and probably the least complex of all the XAI frameworks covered so far, I would say that the framework is not comprehensive enough. Even the visualization provided to analyze the feature contributions appears to be very archaic and can be improved. Since ELI5 is one of the initial XAI frameworks that works with tabular data, images, and text data, it is important to know about it.

In the next section, I will cover the model explainability of H2O AutoML models.

H2O AutoML explainers

Throughout this chapter, we have mostly used scikit-learn-based and TensorFlow-based models. However, when the idea of AutoML was first introduced, the H2O community was one of the earliest adopters of this concept and introduced the AutoML feature for the H2O ML framework: `https://docs.h2o.ai/h2o/latest-stable/h2o-docs/automl.html`.

Interestingly, H2O AutoML is very widely used in the industry, especially for high-volume datasets. Unfortunately, there are very few model explainability frameworks such as DALEX that are compatible with H2O models. H2O models have a good usage in both R and Python, and with the AutoML feature, this framework promises to spin up trained and tuned models to give the best performance in a very short time and with less effort. So, that's why I feel it is important to mention the H2O AutoML explainer in this chapter. This framework does have a built-in implementation of model explainability methods for explaining the predictions of an AutoML model (`https://docs.h2o.ai/h2o/latest-stable/h2o-docs/automl.html`). Next, let's dive deeper into H2O explainers.

Explainability with H2O explainers

H2O explainers are only supported for H2O models. They can be used to provide both global and local explanations. The following list shows the supported methods to provide explanations in H2O:

- Model performance comparison (this is particularly useful for AutoML models that try different algorithms on the same dataset)

- Variable or feature importance (this is for both global and local explanations)

- Model correlation heatmaps

- TreeSHAP-based explanations (this is only for tree models)

- PDPs (this is for both global and local explanations)

- Individual conditional expectation plots, which are also referred to as what-if analysis plots (for both global and local explanations)

You can find out more about H2O explainers at `https://docs.h2o.ai/h2o/latest-stable/h2o-docs/explain.html`. The complete tutorial example is provided at `https://github.com/PacktPublishing/Applied-Machine-Learning-Explainability-Techniques/blob/main/Chapter09/H2o_AutoML_explain_example.ipynb`. In this example, I have demonstrated how to use H2O AutoML for predicting the league position of top football clubs based on the quality of their players using the FIFA Club Position Prediction dataset (`https://github.com/PacktPublishing/Applied-Machine-Learning-Explainability-Techniques/tree/main/Chapter09/datasets/FIFA_Club_Position`). It is the same dataset that we used for the DALEX and Explainerdashboard tutorials.

To install the H2O module, you can use the pip installer:

```
pip install h2o
```

Since we have already covered the steps of data preparation and transformation in the previous tutorials, I will skip those steps here. But please do refer to the tutorial notebook for executing the end-to-end example.

H2O models are not compatible with a pandas DataFrame. So, you will need to convert a pandas DataFrame into an H2O DataFrame. Let's see the lines of code for training the H2O AutoML module:

```
import h2o
from h2o.automl import H2OAutoML
# Start the H2O cluster (locally) - Don't forget this step
h2o.init()
aml = H2OAutoML(max_models=20, seed=1)
train = x_train.copy()
valid = x_valid.copy()
train["position"] = y_train
valid["position"] = y_valid
x = list(train.columns)
y = "position"

training_frame = h2o.H2OFrame(train)
validation_frame=h2o.H2OFrame(valid)

# training the automl model
aml.train(x=x, y=y, training_frame=training_frame,
          validation_frame=validation_frame)
```

Once the AutoML training process is complete, we can get the best model and store it as a variable for future usage:

```
model = aml.get _ best _ model()
```

For the model explainability part, we just need to use the `explain` method from an AutoML model object:

```
aml.explain(validation _ frame)
```

This automatically creates a wide range of supported XAI methods and generates visualizations to interpret the model. At the time of writing, the H2O explainability feature is newly released and is in the experimental phase. If you would like to give any feedback or find any bugs, please raise a ticket request on the H2O JIRA issue tracker (`https://0xdata.atlassian.net/projects/PUBDEV`).

With that, I have covered all the popular XAI frameworks apart from *LIME*, *SHAP*, and *TCAV* that are commonly used or have a high potential for explaining ML models. In the next section, I will give a quick comparison guide to compare all seven frameworks covered in this chapter.

Quick comparison guide

In this chapter, we discussed the different types of XAI frameworks available in Python. Of course, no one framework is absolutely perfect and can be used for all scenarios. Throughout the sections, I did mention the pros and cons of each framework, but I believe it will be really handy if you have a quick comparison guide to decide on your choice of XAI framework, considering your given problem.

The following table illustrates a quick comparison guide for the seven XAI frameworks covered in this chapter. I have tried to compare these based on the different dimensions of explainability, their compatibility with various ML models, a qualitative assessment of human-friendly explanations, the robustness of the explanations produced, a qualitative assessment of scalability, and how fast the particular framework can be adopted in production-level systems:

	DALEX	Explainerdashboard	InterpretML	ALIBI	DiCE	ELI5	H2O AutoML Explainer
Data-Centric Explainability	Yes, but not for comprehension	No support	Yes, but very limited option	No support	No support	No support	No support
Model-Centric Explainability	- Supports local and global explanations - Compares model performance and fairness - Supports interactive visualizations	- Supports local and global explanations - Compares model performance - Supports interactive visualizations	- Supports local and global explanations - Compares model performance - Supports interactive visualizations - But limited explainability methods	- Supports local and global explanations but with limited methods	- Supports local and global explanations but only with counterfactuals	- Supports local and global explanations but only with weight-based feature contributions	- Supports local and global explanations - Compares model performance
Outcome-Centric Explainability	Supports prediction-level explanations using What-If analysis and interactive breakdown plots	Supports prediction-level explanations using What-If analysis	No support	No support	No support	No support	Limited support using ICE plots
Compatibility	Highly compatible with scikit-learn, TensorFlow, PyTorch, and H2O models	Only compatible with scikit-learn model formats	Only compatible with scikit-learn model formats	Compatible with scikit-learn and TensorFlow models	Compatible with scikit-learn and TensorFlow, and PyTorch models	Only compatible with scikit-learn models	Only compatible with H2O models
Human-Friendly Explanation	- High for technical users, especially with the interactive dashboard - Low for non-technical users	- High for technical users, due to interactive dashboards - Low for non-technical users	- Not very human friendly in spite of consolidated dashboard	- Not very human friendly	- CFEs are very close to ideal human-friendly explanations but lack interactive visualizations	- Not very human-friendly	- Useful for mostly technical users
Robustness	Considers different dimensions of explanability, so quite robust	Robustness is high as it provides comprehensive model-centric explanations	Not very robust	Not very robust	CFEs can have a lot of randomness and non-feasible suggestions, so robustness is low	Not very robust	Robustness is high as it provides comprehensive model-centric explanations
Scalability	Seems to be scalable and implemented in a production system	Seems to be scalable and implemented in a production system	Not very scalable	Not very scalable	Seems to be scalable and implemented in a production system	Not very scalable	Scalable for H2O models

Figure 9.24 – A quick comparison guide of the popular XAI frameworks covered in this chapter

This brings us to the end of this chapter. Next, let me provide a summary of the main topics of discussion for this chapter.

Summary

In this chapter, we covered the seven popular XAI frameworks that are available in Python: the *DALEX, Explainerdashboard, InterpretML, ALIBI, DiCE, ELI5*, and *H2O AutoML explainers*. We have discussed the supported explanation methods for each of the frameworks, the practical application of each, and the various pros and cons. So, we did cover a lot in this chapter! I also provided a quick comparison guide to help you decide which framework you should go for. This also brings us to the end of *Part 2* of this book, which gave you practical exposure to using XAI Python frameworks for problem-solving.

Section 3 of this book is targeted mainly at the researchers and experts who share the same passion as I do: *bringing AI closer to end users*. So, in the next chapter, we will discuss the best practices of XAI that are recommended for designing human-friendly AI systems.

References

For additional information, please refer to the following resources:

- The DALEX GitHub project: `https://github.com/ModelOriented/DALEX`
- The Explainerdashboard GitHub project: `https://github.com/oegedijk/explainerdashboard`
- The InterpretML GitHub project: `https://github.com/interpretml/interpret`
- The ALIBI GitHub project: `https://github.com/SeldonIO/alibi`
- The DiCE GitHub project: `https://github.com/interpretml/DiCE`
- The official ELI5 documentation: `https://eli5.readthedocs.io/en/latest/overview.html`
- Model Explainability using H2O: `https://docs.h2o.ai/h2o/latest-stable/h2o-docs/explain.html#`

Section 3 – Taking XAI to the Next Level

This section covers the guidelines and the best practices for applying **Explainable AI (XAI)** to industrial problems. You will learn about some of the open challenges in XAI that could be interesting research problems. User-centric design principles for designing explainable AI/ML systems are also covered in this section, which could potentially help in bringing AI closer to end users and eventually increase the adoption of AI solutions.

This section comprises the following chapters:

- *Chapter 10, XAI Industry Best Practices*
- *Chapter 11, End User-Centered Artificial Intelligence*

10
XAI Industry Best Practices

In the first section of this book, we discussed various concepts related to **Explainable AI** (**XAI**). These concepts were established through years of research, considering various application domains of **artificial intelligence** (**AI**). However, the need for XAI for industrial applications has been felt very recently as AI adoption in industrial use cases is increasing. Unfortunately, the general awareness of XAI for industrial use cases is still lacking due to certain challenges and gaps in how to implement human-friendly explainability methods.

In *Section 2, Practical Problem Solving*, we covered many XAI Python frameworks that are popularly used for interpreting the working of **machine learning** (**ML**) models. However, only understanding how to apply the XAI Python frameworks in practice is not sufficient for industrial problems. Industrial problems require solutions that are scalable and sustainable. So, it is very important for us to discuss the best practices of XAI for scalable and sustainable AI solutions used for industrial problems.

Over the years, XAI has evolved a lot. From being a topic of academic research, XAI is now a powerful tool in the toolkit of AI and ML industrial practitioners. However, XAI has many open challenges on which the research community is still working to bring AI closer to end users. So, we will discuss the existing challenges of XAI and the general recommendations for designing an explainable ML system while considering the open challenges. Also, the quality of AI/ML systems is as good as the quality of the underlying data. Therefore, we will also focus on the importance of adopting a data-first approach for model explainability.

The XAI research community believes that XAI is a multi-disciplinary perspective that should be centered around the end user. So, we will discuss the concept of **interactive machine learning** (**IML**) to create high user engagement for industrial AI systems. Finally, we will cover the importance of providing actionable suggestions and insights using AI/ML as an approach to decipher the complex nature of AI models, thereby making AI explainable and increasing users' trust.

Unlike the previous chapters, in this chapter, we will not focus on the practical applications or learn a new XAI framework. Instead, our goal is to understand the best practices of XAI for industrial use cases. So, in this chapter, we are going to discuss the following topics:

- Open challenges of XAI
- Guidelines for designing explainable ML systems
- Adopting a data-first approach for explainability
- Emphasizing IML for explainability
- Emphasizing prescriptive insights for explainability

So, let's find out more about these topics next.

Open challenges of XAI

As briefly discussed, there have been some significant advances in the field of XAI. XAI is no longer just a topic of academic research; the availability of XAI frameworks has made XAI an essential tool for industrial practitioners. But are these frameworks sufficient to increase AI adoption? Unfortunately, the answer is no. XAI is yet to mature further as there are certain open challenges that, once resolved, can significantly bridge the gap between AI and the end user. Let's discuss these open challenges next:

- *Shifting focus between the model developer and the end user*: After exploring many XAI frameworks throughout this book, you might have also felt that the explainability provided by most of the frameworks requires technical knowledge of ML, mathematics, or statistics to truly understand the working of the model. This is because the explainability methods or algorithms were primarily designed for ML experts or model developers.

  As more and more end users start utilizing AI models and systems, the need for non-technical human-friendly explanations is growing. So, for industrial applications, dynamically shifting the focus between the model developer and the end user is a challenge.

  To a non-technical end user, a simple explanation method such as feature importance visualization can become really complicated unless explicit information is provided. In order to mitigate this challenge, the general recommendation is to design user-centric AI systems. Similar to any software application or system, the user should be involved in the development process early on to understand their requirements and include their expertise while designing the application and not post-production of the application.

- *Lack of stakeholder participation*: From the previous point, although the recommended action is to involve the end users early on in the development process of the AI system, onboarding a stakeholder in the development process can also be a challenge. For most industrial use cases, AI solutions are developed in isolation without involving the final stakeholder(s). Following the design principles from the field of **Human-Computer Interaction** (**HCI**), the user should be involved in the loop during the development process.

Considering high stake domains such as healthcare, finance, legal and regulatory, getting stakeholders and domain experts can be an extremely tedious and expensive process. The stakeholder's availability can be a challenge. Their interest or motivation to participate in the development process can be low even with necessary incentives and compensation. Due to these difficulties in onboarding end users into the development process, designing a user-centric AI system is difficult.

The most recommended action to tackle this challenge is through a collaboration between industry and academia. Usually, academic institutions such as medical schools, law schools, or other universities have broader access to real participants or students who belong to the respective fields and can be *pseudo* participants.

The following diagram illustrates how XAI is a multi-disciplinary perspective:

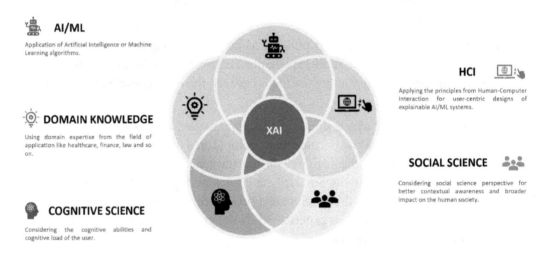

AI/ML
Application of Artificial Intelligence or Machine Learning algorithms.

DOMAIN KNOWLEDGE
Using domain expertise from the field of application like healthcare, finance, law and so on.

COGNITIVE SCIENCE
Considering the cognitive abilities and cognitive load of the user.

HCI
Applying the principles from Human-Computer Interaction for user-centric designs of explainable AI/ML systems.

SOCIAL SCIENCE
Considering social science perspective for better contextual awareness and broader impact on the human society.

XAI

Figure 10.1 – XAI is a multi-disciplinary perspective

- *Application-specific challenges*: Different application domains need explainability of different types. For example, in an AI-based loan approval system, influence-based or example-based feature explanations can be really helpful. However, for an application to detect COVID-19 infections from X-ray images, highlighting or localizing the region of the infection can be more helpful. So, each application can have its own requirement and definition of explainability and, thus, any general XAI framework might not be very effective.

- *Lack of quantitative evaluation metrics*: The quantitative evaluation of explanation methods has been an important research topic. Unfortunately, there is still no tool or framework that exists that can quantitatively evaluate the quality of explanation methods. This is mostly because many diverse AI algorithms are at work on different types of data. Consequently, there are many definitions of model explainability and many approaches for XAI. So, it is very hard to generalize quantitative evaluation metrics that can work with all of the different explanation methods.

  Currently, qualitative evaluation methods such as *Trust, Usefulness, Actionability, Coherence with prior beliefs, Impact*, and more are used. To learn more about these metrics, take a look at *Understanding Machines: Explainable AI* from *Accenture Labs*, which is available at `https://www.accenture.com/_acnmedia/pdf-85/accenture-understanding-machines-explainable-ai.pdf`. Additionally, take a look at *Explanation in Artificial Intelligence: Insights from the Social Sciences* from *Tim Miller*, which is available at `https://arxiv.org/pdf/1706.07269.pdf`.

  The qualitative evaluation methods are, indeed, user-centric and use the principles of HCI to collect feedback from the end user, but usually, quantitative metrics are more useful when comparing different methods. However, I am hopeful that tools such as *Quantus* (`https://github.com/understandable-machine-intelligence-lab/Quantus`), which is used to evaluate explanation methods for neural networks, will mature significantly in a few years and it will be easier to evaluate explanation methods.

- *Lack of actionable explanations*: Most explanation methods don't provide actionable insights to the end user. So, designing explainable AI/ML systems that can provide actionable explanations can be challenging. Counterfactual explanations, what-if analysis, and interactive visualization-based explanations are the only explanation methods that allow the user to observe the change in outcome when the input features are altered. I would recommend increasing the usage of these actionable explanation methods to develop explainable AI/ML systems.

- *Lack of contextual explanations*: Any ML algorithm that is deployed in production depends on the specific use case and the underlying data. Due to this, there is always a trade-off between explainability, model performance, fairness, and privacy. So, understanding the context of explainability is an existing challenge that any general XAI framework cannot provide accurately. So, the recommendation to mitigate this challenge is to design personalized explainable ML systems for a specific use case rather than a generalized implementation.

If you want to explore more in this area, you can take a look at *Verma et al.'s* work, *Pitfalls of Explainable ML: An Industry Perspective* (`https://arxiv.org/abs/2106.07758`), to learn more about the typical challenges of XAI. All of these open challenges are interesting research problems that you can explore to help the research community progress in this field. Now that we have discussed the open challenges of XAI, next, let's discuss the guidelines for designing explainable ML systems for industrial use cases, considering the open challenges.

Guidelines for designing explainable ML systems

In this section, we will discuss the recommended guidelines for designing an explainable ML system from an industry perspective while considering the open challenges of XAI, as discussed in the previous section. All of these guidelines have been carefully collated from various publications, conference keynotes, and panel discussions from various experts in the field of XAI, ML, and software systems. It is true that every ML and AI problem is unique in its own way, and so, it is hard to generalize any recommendations. But many AI organizations have adopted the following list of guidelines for designing explainable and user-friendly ML systems:

- *Identify the target audience of XAI and their usability context*: The definition of explainability depends on the user using the AI system. *Arrieta et al.*, in their work *Explainable Artificial Intelligence (XAI): Concepts, Taxonomies, Opportunities, and Challenges toward Responsible AI*, have highlighted the importance of identifying the target audience of XAI when designing explainable AI systems.

  An AI system can have different audiences such as technical stakeholders (that is, data scientists, ML experts, product owners, and developers), business stakeholders (that is, managers and executive leaders), domain experts (that is, doctors, lawyers, insurance agents, and more), legal and regulatory agencies, and non-technical end users. Every audience might have a different need for explainability, so accordingly, the explanation methods should try to address the best needs of the audience.

As a preliminary step, identifying the target audience of the explainable system along with the situation or context in which they are going to use the system helps a lot in the design process. For example, for medical experts relying on ML models to predict the risk of diabetes, the choice of explanation methods depends on their actual needs. If their need is to suggest actions to improve the health conditions of diabetic patients, then counterfactual examples can be really useful. However, if their purpose is to find out the factors that are leading to the increase in the risk of diabetes, then feature-based explanations methods are more relevant.

Figure 10.2 illustrates the various target audience of explainable AI systems:

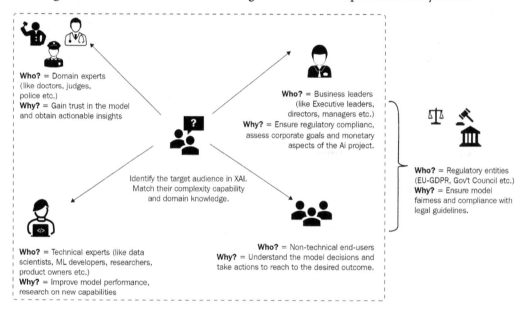

Figure 10.2 – Identifying the target audience of XAI

- *Shortlisting the XAI techniques based on the user's needs*: Once the target audience and their usability context have been identified, along with the necessary technical details about the type of the dataset (for instance, tabular, images, or textual) and the ML algorithm used for training the model, shortlisting a list of possible explanation methods, as covered in *Chapter 2, Model Explainability Methods*, is very important.

These shortlisted explanation methods should fit in with the software system that the target users will use to interact with the AI models. This means the explanation techniques should be well integrated with the software applications or interfaces and should even be considered during the design process of the software for a consistent user experience.

- *Human-centered XAI: An iterative process of translating and evaluating XAI in specific domains involving the end user*. Similar to the design life cycle of a software system using HCI, XAI is also an iterative process. It should be human-centered and should be evaluated continuously to assess the impact. In the *User-centric system design using XAI* section of *Chapter 11, End User-Centered Artificial Intelligence*, I have included other important aspects to consider for a human-centered XAI design process.

- *The importance of the feedback loop in XAI*: All explainable AI systems should have the option to capture the end user's feedback to assess the impact, relevance, effectiveness, and trust of the explanations provided by the system. It is never possible to consider all edge cases and all preferences of the end users during the design and the initial development process. But using the feedback loop, developers can collect specific feedback about the explanation methods and modify them if needed.

- *The importance of scalability in the design process*: Similar to serving ML models for production systems, explainability should also be served in modular and scalable approaches. The best way to serve model explanations is by designing **scalable web APIs** to be deployed in centralized cloud servers. So, when XAI is implemented in practice, do make sure that the explanations are being served through web APIs so that they can be easily integrated with any software interface or application.

- *Toggling between the data, the interface, and actionable insights*: It has been observed by many experts that, for end users, their satisfaction with the model explanation method is a trade-off between how well the explanation is being connected to the underlying dataset (or their prior beliefs), how the users are able to interact with the ML system to gain more confidence in it, and how well the explanations encourage them to take actions to get their desired output. **Data-centric XAI, IML, and actionable explanations** are broader research topics that should be considered when designing the explainable AI system for industrial use cases.

So, we have learned about the open challenges of XAI and discussed the design guidelines considering the open challenges. We now have a fair idea of what to consider when designing explainable ML systems. Next, let's elaborate on the last recommended guideline in the upcoming sections to carefully understand why it is important. Let's start our discussion with the importance of using a data-centric approach for explainability.

Adopting a data-first approach for explainability

In *Chapter 3, Data-Centric Approaches*, we discussed the importance and various techniques of **Data-Centric XAI**. Now, in this section, we will elaborate on how adopting a data-first approach for explainability helps in gaining users' trust in industrial use cases.

Data-centric AI is based on the fundamental idea that *the quality of the ML model is as good as the quality of the underlying dataset used for training the model*. For industrial use cases, dealing with poor-quality datasets is a major challenge for most data scientists. Unfortunately, data quality is often ignored as data scientists and ML experts are expected to cast their *magic* of ML to build models that are close to 100% accurate. Consequently, ML experts simply try to follow **model-centric approaches** such as tuning hyperparameters or using complex algorithms to boost model performance. Even if the model performance increases slightly, with the increase in complexity, explainability decreases. The lack of explainability increases the skepticism of the business stakeholders. Also, issues relating to data quality such as the presence of *data anomalies*, *data leakage*, *data drift*, and other issues, as discussed in *Chapter 3, Data-Centric Approaches*, significantly increase. *In that case, what do we do?*

The answer is to adopt a data-centric approach to explain the ML process. Using data-centric explainability methods such as **exploratory data analysis** (**EDA**), we can extract insights about the dataset such as any interesting patterns, correlations, monotonicity, or trends from the features used in the dataset. EDA and data analysis between the training data and the inference data also helps you to identify data quality issues. If there are issues in the dataset, it is always recommended that you inform the business stakeholder about the limitations of poor data quality and set the expectations correctly about the model performance. So, even if the model predictions are not correct, the business stakeholder will understand the limitations instead of doubting the ML system.

But *why don't we try out the other XAI frameworks and methods covered throughout this book? How would adopting a data-first approach for explainability help?* Well, you can and you should try out other relevant XAI methods if applicable, but data-centric explainability is always easier to explain to a non-technical user. Especially, with the *data profiling method*, as discussed in *Chapter 3, Data-Centric Approaches*, we can identify the range of values of features present in the dataset for each category (if there is a classification problem) or each bin of the prediction variable (if there is a regression problem) and compare the model predictions with the profiled values. Simple comparisons with the profiled values are easier to understand as compared to complicated mathematical concepts such as *Shapley values* or other algorithms used in XAI frameworks.

Another reason why data-centric approaches are preferable is because of the user's trust in historical data. Generally, it is observed that most stakeholders have more trust in historical data as compared to AI models. For example, in the spring season, if an AI weather forecasting model predicts the occurrence of snowfall, most end users would be hesitant to trust the prediction. That's because spring is always associated with sunshine and flowers blooming due to the observations throughout the world over many years. But if the model also indicates the occurrence of snowfall in the last few years during the same time or even indicates that there was snowfall in close proximity in the last few days, the user's trust would be greater. So, it is recommended that you, first, explore data-centric explainability and then look at other explainability methods for any industrial ML problems.

The following diagram illustrates how data-centric XAI can be very close to the natural ways of providing explainability, thereby improving the ease of understanding:

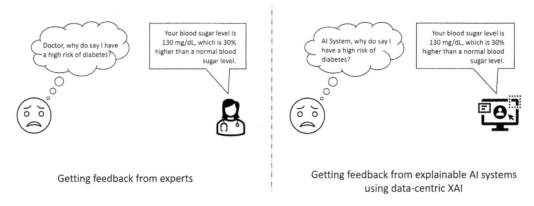

Figure 10.3 – The importance of a data-centric approach for explainability

Next, we will discuss interactive ML to boost the end user's trust.

Emphasizing IML for explainability

IML is the paradigm of designing intelligent user interfaces to facilitate ML and AI algorithms with the help of user interactions. Using IML to steer the usage of ML systems to increase the trust of the end user has been an important research topic for the AI and HCI research community over the last few years. Many works of research literature recommend using IML to increase user engagement for AI systems. *Recent Research Advances on Interactive Machine Learning* by *Jiang et al.* (https://arxiv.org/abs/1811.04548) talks about some of the significant progress that has been made in the field of IML and how it is closely associated with the increasing trust and transparency of ML algorithms.

IML is another interesting approach that is used by the XAI community to explain ML models. Even in frameworks such as *DALEX* and *Explainerdashboards*, as covered in *Chapter 9, Other Popular XAI Frameworks*, providing interactive dashboards and web interfaces that end users can interact with to explore the data, model, and predictions are considered as a way for model explainability. IML helps the user in the following ways:

- Explore the dataset through graphs and visuals, thereby making it easier for the user to observe and remember key insights from the data.

- Gain more confidence about the ML systems, as the intelligent user interfaces allow the user to make changes and observe the outcome. It makes it easier for the user to figure out how the model behaves while considering any changes in input.

- Typically, what-if analysis and local explainability are improved when interactive interfaces are provided.

- IML gives more control to the user to explore the system, and IML usually considers a user-centric design process for providing customized interfaces tailor-made for a specific use case.

In short, IML improves the user experience and, thus, helps to boost the adoption of AI models. I would strongly recommend using interactive user interfaces as part of explainable ML systems along with serving model explainability using modularized web APIs. You can read the following article to find out more about the usefulness of IML for business problems: `https://hub.packtpub.com/what-is-interactive-machine-learning/`.

The following diagram illustrates the difference between conventional ML and IML:

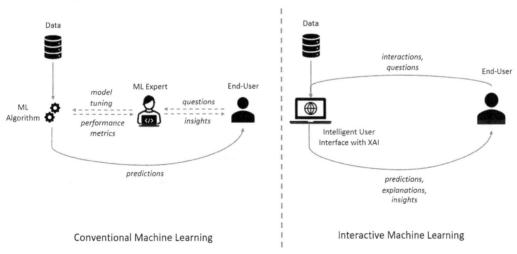

Figure 10. 4 – Comparing conventional ML with IML

As you can see from the preceding diagram, using IML, the end user can directly interact with the intelligent user interface to get predictions, explanations, and insights. Next, let's discuss the importance of prescriptive insights for explainable ML systems.

Emphasizing prescriptive insights for explainability

Prescriptive insight is a popular jargon used in data analysis. It means providing actionable recommendations derived from the dataset to achieve the desired outcome. It is often considered to be a catalyst in the entire process of data-driven decision-making. In the context of XAI, explanation methods such as *counterfactual examples*, *data-centric XAI*, and *what-if analysis* are prominently used for providing actionable suggestions to the user.

Along with counterfactuals, the concept of **actionable recourse in ML** is also used for generating prescriptive insights. **Actionable recourse** is the ability of a user to alter the prediction of an ML model by modifying the features that are actionable. But *how is it different from counterfactuals?* Actionable recourse can be considered to be an extension of the idea of counterfactual examples, which uses actionable features instead of all the features present in the dataset.

Now, *what do we mean by actionable features?* Considering a practical scenario, it is not feasible for us to change all the features present in a dataset in any direction to reach the desired outcome. For example, features such as *age*, *gender*, and *race* cannot be changed in any direction to obtain the desired output. Unfortunately, algorithms used for generating counterfactual examples do not consider the practical feasibility of changing a feature.

Let's suppose that an ML model is being used to estimate the risk of diabetes. For a diabetic patient, if we want to use counterfactual examples to recommend how to reduce the risk of diabetes, it is not practically feasible for the patient to reduce their age by 10 years or change their gender to decrease the risk. So, these are non-actionable features. Even though theoretically altering these features can change the model prediction, it is not practical to change these features. Therefore, the concept of actionable recourse is more like a controlled counterfactual generation process that is applied to actionable features and considers a practically feasible boundary condition for the feature values.

To generate prescriptive insights, I would recommend that you use actionable recourse as it considers the practical feasibility and difficulty of altering a feature value to get the desired outcome. You can find more about actionable recourse from *Ustun et al.'s* work, *Actionable Recourse in Linear Classification* (https://arxiv.org/abs/1809.06514), along with their GitHub project at https://github.com/ustunb/actionable-recourse.

But *are prescriptive insights really necessary in XAI?* Well, the answer is *yes!* The following list of reasons explains why prescriptive insights are important in XAI:

- Prescriptive insights are actions suggested to the user to get the desired result. In most industrial use cases, explainability is incomplete if the user is unaware of how to reach their desired outcome.

- Generating prescriptive insights is a proactive method for explaining the working of ML models. That's because it allows the user to take necessary proactive actions rather than trusting the passive explanations provided to them.

- It increases the user's faith in the system by giving a sense of control over the system. Using actionable explanations, the user is empowered to alter the model prediction.

- It increases the ability of business stakeholders for making data-driven decisions for the organization.

These are the main reasons why you should always consider generating explanations that are actionable when designing explainable AI systems for industrial problems. *Figure 10.5* illustrates how prescriptive insights using XAI can provide actionable recommendations for the user to get their desired outcome:

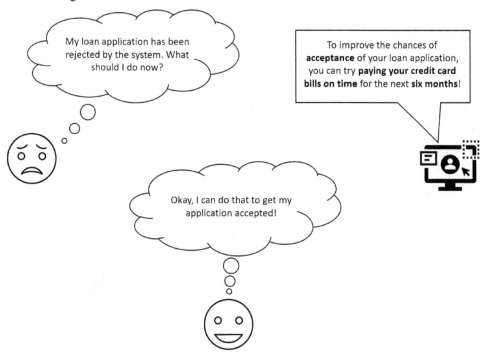

Figure 10.5 – The importance of prescriptive insights for explainability

With this, we have arrived at the end of this chapter. Let's summarize the topics discussed next.

Summary

This chapter focused on the best practices for designing explainable AI systems for industrial problems. In this chapter, we discussed the open challenges of XAI and the necessary design guidelines for explainable ML systems, considering the open challenges. We also highlighted the importance of considering data-centric approaches of explainability, IML, and prescriptive insights for designing explainable AI/ML systems.

If you are a technical expert, architect, or business leader responsible for using AI to solve industrial problems, this chapter has helped you to learn some of the most important guidelines for designing explainable AI/ML systems considering the open challenges in XAI. If you are a researcher in the field of AI or HCI, some of the open challenges discussed in the chapter could be interesting research topics to consider. Finding solutions to these challenges can lead to significant progress in the field of XAI.

In the next chapter, we will cover the principles of **End User-Centered Artificial Intelligence** to bridge the AI-end user gap.

References

For additional information about the topics covered in this chapter, please refer to the following resources:

- *Pitfalls of Explainable ML: An Industry Perspective*: https://arxiv.org/abs/2106.07758

- The Quantus framework in GitHub: https://github.com/understandable-machine-intelligence-lab/Quantus

- *Explanation in Artificial Intelligence: Insights from the Social Sciences*: https://arxiv.org/pdf/1706.07269.pdf

- *Quantus: An Explainable AI Toolkit for Responsible Evaluation of Neural Network Explanations*: https://arxiv.org/abs/2202.06861

- *Understanding Machines: Explainable AI* from *Accenture Labs*: https://www.accenture.com/_acnmedia/pdf-85/accenture-understanding-machines-explainable-ai.pdf

- *Actionable Recourse in Linear Classification* by *Ustun et al*: `https://arxiv.org/abs/1809.06514`

- Actionable recourse in ML: `https://github.com/ustunb/actionable-recourse`

- *Questioning the AI: Informing Design Practices for Explainable AI User Experiences* by *Liao et al*: `https://dl.acm.org/doi/10.1145/3313831.3376590`

- *Advances and Open Questions in Explainable AI (XAI): A practical perspective from an HCI researcher* by *Q. Vera Liao*: `http://qveraliao.com/aaai_panel.pdf`

11

End User-Centered Artificial Intelligence

Over the last 10 chapters of this book, we have traveled over the entire landscape of **Explainable AI** (**XAI**), covering different types of explainability methods used in practice for different dimensions of explainability (*data*, *model*, *outcome*, and the *end users*). XAI is an active field of research that I think is yet to reach its full potential. But the field is growing rapidly, along with the broader domain of AI, and we will witness many new algorithms, approaches, and tools being developed in the future. Most likely, the new methods and tools of XAI will be better than the existing ones and will be able to tackle some of the *open challenges of XAI* discussed in *Chapter 10, XAI Industry Best Practices*. Unfortunately, we cannot extend the scope of this book to cover all possible approaches to XAI. However, the goal of this book is to provide a blend of conceptual understanding of the field with the required practical skills so that it is a useful starting point for beginners, and even add to the knowledge of experts for an applied knowledge of XAI.

In the previous chapter, we discussed the recommended practices for implementing an explainable **Machine Learning** (**ML**) system from the industry perspective. We also discussed the existing challenges of XAI and some recommended ways to mitigate the challenges. Considering these existing challenges, in this chapter, we will focus on the ideology of **End User-Centered Artificial Intelligence** (**ENDURANCE**). This is a term that is often used to refer to sustainable and scalable AI solutions that are built, keeping the user in the center. It is recommended that you read the previous chapter before starting this chapter for a better understanding. ENDURANCE is neither a new algorithm nor a new, sophisticated tool for XAI. Instead, it is a practice; it is a methodical discipline to bridge the AI-end user gaps.

This chapter will be particularly useful for researchers from the field of AI and **Human-Computer Interaction** (**HCI**) who view XAI from a *multidisciplinary perspective*. It is also useful for business leaders who want to drive problem solving using AI, considering a seamless **User Experience** (**UX**). For AI developers and thought leaders, this chapter will help you to design your AI solutions keeping the end user in the center and promoting AI adoption.

This chapter focuses on the following main topics:

- User-centered XAI/ML systems
- Rapid XAI prototyping using EUCA
- Efforts toward increasing user acceptance of AI/ML systems using XAI
- Providing delightful UX
- What's next in XAI?

Let's proceed with the first topic of discussion in the next section.

User-centered XAI/ML systems

For most industrial problems, AI solutions are developed in isolation and users are only introduced in the final stages of the development process after a minimum viable solution is ready. With this conventional approach, it is often found that product leads or product managers tend to focus on projecting the solution from the development team's perspective to meet the goals of the users. Well, this approach is absolutely fine, and it might work really well for certain use cases that require the technical team to develop through abstraction. However, if the users are not involved in the early stages of the implementation process, it has been often found that the users are reluctant to adopt the solution. So, the ENDURANCE ideology is focused on developing solutions by involving final users right from the design phase of the solution.

The ENDURANCE ideology focuses on the principles of HCI and emphasizes the importance of *distributed cognition* of the user. With this ideology, the entire solution comprising the **User Interface (UI)**, *AI algorithms*, *underlying dataset*, *XAI component*, and *end user's experience* is considered collectively as a *system*, rather than considering the individual components in isolation. This ensures that explainability is baked into the system instead of being offered as an add-on service for the user. From what I have observed, most industrial AI solutions are developed in isolation as a separate component and then added to the main software system as an *add-on* or *premium feature*. Similarly, the XAI component is also considered an add-on feature after being developed in isolation. Consequently, the seamless UX can get hampered, and the main benefits of the AI solution and the XAI component may not be realized to their full potential. This is why we should focus on the design and development of the entire user-centric XAI/ML system.

Next, let's discuss the various aspects of end user-centric XAI that we should consider while designing the solution.

Different aspects of end user-centric XAI

In this section, we will discuss the different principles of human factors that should be integrated while designing the XAI system using the ENDURANCE ideology for bridging the AI and end user gap.

Goal relevance

The primary questions that the field of HCI tries to address are *Who are the users?* and *What are their needs?* Or in other words, it tries to understand the *goal relevance* of the solution for the user. If the solution provided is not effectively solving the problem by meeting the needs of the users without introducing other challenges, it is not relevant. Not considering the goal relevance is probably one of the main reasons why the majority of AI solutions are either scrapped or adopted with a lot of skepticism.

The recommended approach to evaluate goal relevance is by checking whether the users can achieve their goals without the introduction of other challenges. Along with goal relevance, I often recommend assessing the impact of the solution. The impact of the solution can be qualitatively measured by taking the user's feedback when the solution is absent.

Connecting the user needs with the strengths of AI

As discussed before, in most industrial use cases, XAI is used in isolation to provide explainability without considering the user needs. Instead, using the ideology of ENDURANCE, XAI should connect the user needs with the strength of the AI algorithm. Once the user needs are identified, *translate the user needs into data needs and model needs*. If the underlying dataset is not sufficient to meet all the user needs, use **data-centric XAI** to communicate the limitations of the dataset to the user. If the model needs are identified, use XAI to interpret the working of the model, and tune accordingly to meet the needs of the user.

But this process can be challenging as it involves identifying the existing *mental model* of the user. With the introduction of AI and XAI, the existing workflow should not get disrupted.

Moreover, it is also recommended that using XAI, you try to explain whether the AI solution is adding any unique value. But design the explainability methods to justify the advantages and not the underlying technology used. For example, if the system conveys to the user that complex deep learning algorithms are used to predict the outcome, it does not increase the confidence of the user. Instead, if the system conveys that the intelligent solution helps the user to reach their goal five times faster than the conventional approach, the user will agree to adopt the solution.

User interface – a medium to connect users with the AI solution

Considering the conventional approaches, most AI practitioners are focused only on developing accurate AI models giving much less focus to the user's interaction with the model. Generally, the user's interaction with the AI component is decided by the software engineering teams; unfortunately, in most organizations, the data science and AI teams work in silos. But it is the UI that controls the level of visibility, explainability, or interpretability of the AI models and plays a vital role in influencing the user's trust in the system.

In *Chapter 10*, *XAI Industry Best Practices*, while discussing **Interactive Machine Learning** (**IML**), we discussed how the user's interaction with the system through the UI gives more confidence to the user about the working of the AI/ML system. Hence, the UI should be in alignment with the AI model and its explainability methods to calibrate the user's trust. You can find out more about calibrating the user's trust using the UI in the People + AI Guidebook from Google PAIR: `https://pair.withgoogle.com/chapter/explainability-trust/`.

Involvement of the end user early in the development process of the solution

Unlike conventional approaches, the user-centric approach recommends involving the final user(s) early in the development process. The end user should in fact be involved from the design phase of the UI of the system, so that the needs of the user are correctly mapped into the interface. Similar to the design and development life cycle of the solution, explainability should also be evolved in an iterative process by taking continuous feedback from the user.

As the ENDURANCE ideology views the XAI/ML system as one solution, the entire solution should have a *design phase, prototype phase, development phase*, and *evaluation phase*. These four phases would collectively form *one iteration of design and development*. Likewise, the entire solution should be matured in several iterations, keeping the user involved in every single phase of each iteration. This process is also in alignment with the *agile methodology* followed in software engineering. Involvement of the user in every phase ensures that useful feedback is collected for evaluating whether the user's needs are being met by the solution. Early involvement also ensures that the users are familiar with the design and working of the new system. Users' familiarity with the system increases the adoption rate of the system.

Connecting feedback with personalization

As discussed in the previous section, the importance of the user's feedback in every phase of the design and development of the solution is inevitable. But sometimes, a general framework of a solution doesn't fulfill all needs of the user.

For example, when using counterfactual examples, it is technically possible to generate an example using all the features used for the prediction. But suppose the user is only interested in changing a specific set of actionable variables. In that case, the controlled counterfactuals should modify only the features that are interesting to the user. It has been found that a tailor-made personalized solution is often more useful to the end user than a generalized solution. So, using the feedback obtained from the user, try to provide a personalized solution meeting the specific pain points of the user.

Contextual and actionable AI

As we previously discussed in *Chapter 10, XAI Industry Best Practices*, explanations should be contextual and actionable. The entire XAI/ML system should also be in alignment with the user's actions and should have context awareness. XAI plays a vital role in connecting AI to the user's action and modifying any AI solution into a contextual AI solution.

Oliver Brdiczka, in his article *Contextual AI: The Next Frontier of Artificial Intelligence* (`https://business.adobe.com/blog/perspectives/contextual-ai-the-next-frontier-of-artificial-intelligence`), defined the following four pillars of contextual AI:

- **Intelligible**: Contextual AI systems should be able to explain its knowledge and working.

- **Adaptive**: Contextual AI systems should be able to adapt to the different needs of the user in a different environment.

- **Customizable**: The users should be able to control or modify the system to meet their needs.

- **Context-aware**: The system should be able to perceive things at the same level as a human.

The following figure shows the four different components of contextual AI:

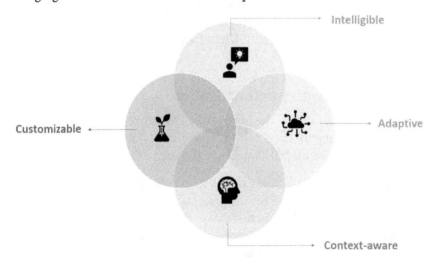

Figure 11.1 – Four components of contextual AI (inspired by https://business.adobe.com/blog/perspectives/contextual-ai-the-next-frontier-of-artificial-intelligence)

So, considering user-centric approaches, the XAI component of XAI/ML systems should provide actionable insights and it should be contextual to further bridge the gap between AI and end users. Now that we have discussed the user-centric approaches to bridge possible gaps between AI and end users, considering the open challenges of XAI discussed in *Chapter 10*, *XAI Industry Best Practices*, let's discuss making rapid XAI prototypes using the **End User-Centric Explainable Artificial Intelligence (EUCA)** framework.

Rapid XAI prototyping using EUCA

In the previous section, we discussed the key ingredients of a user-centered XAI/ML system. In this section, the importance of rapid prototyping in the ENDURANCE ideology will be emphasized. *Rapid prototyping* is a concept that is predominantly adopted in software engineering as software is probably the most malleable thing created by mankind. Building fast prototypes is an approach for collecting useful user feedback early in the development process of a software product. Hence, even for designing user-centered XAI/ML systems, rapid prototyping is very important.

Jin et al., in their research work *EUCA: the End-User-Centered Explainable AI Framework* (`https://arxiv.org/abs/2102.02437`), introduced a toolkit called EUCA. EUCA is a very interesting framework primarily designed by UX researchers, HCI researchers and designers, AI scientists, and developers for building rapid XAI prototypes for non-technical end users. The official GitHub repository for the EUCA framework is available at `https://github.com/weinajin/end-user-xai`. It is strongly recommended to use EUCA to build low-fidelity prototypes and iteratively improve the prototype based on continuous user feedback for XAI/ML systems.

The following important components are offered by this framework:

- 12 explanatory forms for designing human-friendly explanations

- Corresponding XAI algorithms for integrating with functional prototypes

- Associated design templates and examples of their usage

- Suggested prototyping workflows

- Detailed strengths and weaknesses of various explanation methods obtained from their user study findings

- Scientific analysis of diverse explanation needs of end users (such as calibration of trust, detection of bias, and resolution of disagreements with AI)

The following figure illustrates the different types of explanation methods currently supported by the EUCA framework:

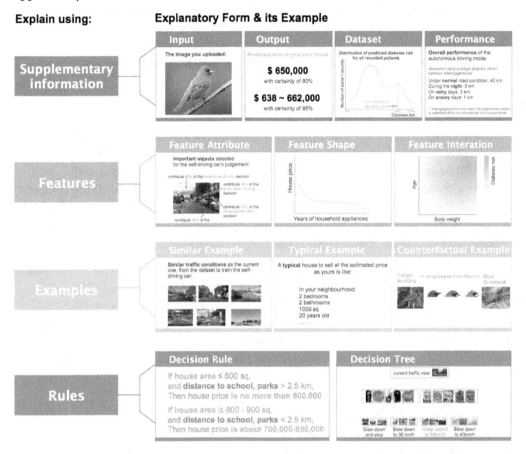

Figure 11.2 – Different types of explanation methods supported in EUCA

This framework is a great starting point and definitely recommended for building rapid XAI prototypes. Next, let's discuss some additional efforts that can be made to increase user acceptance of AI/ML systems.

Efforts toward increasing user acceptance of AI/ML systems using XAI

In this section, we will discuss some recommended practices to increase the acceptance of AI/ML systems using XAI. In most software systems, the **User Acceptance Testing (UAT)** phase is used to determine the *go* or *no-go* for software. Similarly, before the final production phase, more and more organizations prefer doing a robust UAT process for AI/ML systems. But *how important is the explainability of AI algorithms, when doing UAT of AI/ML systems? Can explainability increase the user acceptance of AI?* The short answer is *yes!* Let's go through the following points to understand why:

- **User acceptance is a testimony of the user's trust** – Since XAI can increase the user's trust in AI, it increases the chance of the user's acceptance of the solution. Now, trust is something that cannot just be established during the UAT phase; rather, trust should be established from the beginning and maintained throughout the development process. The capabilities and limitations of the system should be communicated from the beginning to set a clear expectation of what is possible and what is not possible.

- **Risk tolerance estimation as UAT criteria** – It is quite obvious that AI systems cannot be 100% accurate every single time. It is not practically possible to achieve systems that have zero error or zero failure. But as a recommended practice, it is important to document the possible failure points for the system and the consequences of the potential failures of the system are termed **risk**. **Risk tolerance** is the maximum permissible error that the system can make without causing a huge impact. So, during the UAT phase, it is important to define the risks of the solution and have an estimation of the maximum risk tolerance of the user. The system's ability to perform within the risk tolerance should be considered a success criterion for the UAT process.

- **Perform as many user studies as possible before the UAT process** – User studies and qualitative and quantitative analysis of the user's feedback are certain ways by which the impact and trust of the system can be assessed. So, perform multiple user studies before the UAT process and ensure that the users are accepting the prototype solutions before directly moving the system into production.

The preceding approaches are certain ways to increase user acceptance, but ultimately, user acceptance depends on the overall UX. In the next section, we will discuss further the importance of providing a delightful UX.

Providing a delightful UX

In this section, we will focus on the importance of overall UX to promote the adoption of XAI/ML systems. *Aaron Walter*, in his book *Designing for Emotion* (https://abookapart.com/products/designing-for-emotion), mentioned some of the foundational elements of user needs that must be met before higher motivation can influence the behavior of the user. According to his hierarchy of user needs, *pleasurable* or *delightful* UX is at the top of the pyramid. The following figure shows Aaron Walter's hierarchy of user needs:

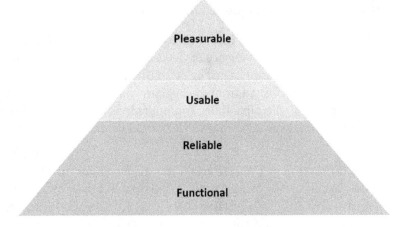

Figure 11.3 – Aaron Walter's hierarchy of user needs

This hierarchy of user needs defines the fundamental needs of the end user that should be fulfilled before any advanced needs of the user are addressed. So, if a system is only *functional*, *reliable*, and *usable*, it is not sufficient for adopting the system unless the overall UX is delightful and enjoyable! Hence, XAI/ML systems should also consider providing a seamless overall experience to truly bridge the AI-end user gap.

This brings us to the end of the last chapter of this book. We will summarize the key topics of discussion in the next section.

Summary

In this chapter, we have primarily discussed using the ideology of ENDURANCE for the design and development of XAI/ML systems. We have discussed the importance of using XAI to steer us toward the main goals of the end user for building XAI/ML systems. Using some of the principles and recommended best practices presented in the chapter, we can bridge the gap between AI and the end user to a great extent!

This also brings us to the end of this book! Congratulations on reaching the end! This book was carefully designed to include conceptual understanding of various XAI concepts and jargon, practical examples to use popular XAI frameworks for applied problem solving, real-life examples and experiences from an industrial perspective, and references to important research literature to further expand your knowledge. This book introduced you to the field of XAI from both the industrial perspective as well as an academic research perspective. The open challenges and the next phases of XAI research topics discussed in this book are important research problems that are being explored by the AI research community.

Even though this book touched on almost every aspect of the field of XAI, clearly there are lots more to explore and unravel. My recommendation is not to restrict yourself to what was offered by this book. Instead, use this book as a reference starting point but explore and apply the knowledge gained from this book to practical use cases and step forward to contribute to the community!

References

Please refer to the following resources to gain additional information:

- *People + AI Guidebook from Google PAIR*: https://pair.withgoogle.com/chapter/explainability-trust/

- *Jin et al., EUCA: the End-User-Centered Explainable AI Framework*: https://arxiv.org/abs/2102.02437

- *EUCA: End-User-Centered Explainable AI Framework GitHub repository*: https://github.com/weinajin/end-user-xai

- *Aaron Walter, Designing for Emotion*: https://abookapart.com/products/designing-for-emotion

- *Oliver Brdiczka, Contextual AI: The Next Frontier of Artificial Intelligence*: https://business.adobe.com/blog/perspectives/contextual-ai-the-next-frontier-of-artificial-intelligence

Index

Symbols

A

Packt.com

Subscribe to our online digital library for full access to over 7,000 books and videos, as well as industry leading tools to help you plan your personal development and advance your career. For more information, please visit our website.

Why subscribe?

- Spend less time learning and more time coding with practical eBooks and Videos from over 4,000 industry professionals
- Improve your learning with Skill Plans built especially for you
- Get a free eBook or video every month
- Fully searchable for easy access to vital information
- Copy and paste, print, and bookmark content

Did you know that Packt offers eBook versions of every book published, with PDF and ePub files available? You can upgrade to the eBook version at packt.com and as a print book customer, you are entitled to a discount on the eBook copy. Get in touch with us at customercare@packtpub.com for more details.

At www.packt.com, you can also read a collection of free technical articles, sign up for a range of free newsletters, and receive exclusive discounts and offers on Packt books and eBooks.

Other Books You May Enjoy

If you enjoyed this book, you may be interested in these other books by Packt:

Hands-On Explainable AI (XAI) with Python

Denis Rothman

ISBN: 9781800208131

- Plan for XAI through the different stages of the machine learning life cycle
- Estimate the strengths and weaknesses of popular open-source XAI applications
- Examine how to detect and handle bias issues in machine learning data
- Review ethics considerations and tools to address common problems in machine learning data
- Share XAI design and visualization best practices
- Integrate explainable AI results using Python models
- Use XAI toolkits for Python in machine learning life cycles to solve business problems

Interpretable Machine Learning with Python

Serg Masís

ISBN: 9781800203907

- Recognize the importance of interpretability in business
- Study models that are intrinsically interpretable such as linear models, decision trees, and Naïve Bayes
- Become well-versed in interpreting models with model-agnostic methods
- Visualize how an image classifier works and what it learns
- Understand how to mitigate the influence of bias in datasets
- Discover how to make models more reliable with adversarial robustness
- Use monotonic constraints to make fairer and safer models

Packt is searching for authors like you

If you're interested in becoming an author for Packt, please visit `authors.packtpub.com` and apply today. We have worked with thousands of developers and tech professionals, just like you, to help them share their insight with the global tech community. You can make a general application, apply for a specific hot topic that we are recruiting an author for, or submit your own idea.

Share Your Thoughts

Now you've finished *Applied Machine Learning Explainability Techniques*, we'd love to hear your thoughts! Scan the QR code below to go straight to the Amazon review page for this book and share your feedback or leave a review on the site that you purchased it from.

https://packt.link/r/1803246154

Your review is important to us and the tech community and will help us make sure we're delivering excellent quality content.

www.ingramcontent.com/pod-product-compliance
Lightning Source LLC
Chambersburg PA
CBHW062110050326
40690CB00016B/3273